A

Since his retirement from university teaching, Michael Cohen has been writing personal essays about his family, about lifelong pursuits such as golf and birding, about newer interests in flying and amateur astronomy, and above all about six decades of reading.

His essays – collected here in *A Place to Read* – have appeared in *Harvard Review*, *Birding*, *The Humanist*, *The Missouri Review*, and *The Kenyon Review*.

He is the author of five books, including an introductory poetry text, *The Poem in Question* (Harcourt Brace, 1983) and an award-winning book on Shakespeare's *Hamlet* (Georgia, 1989).

Michael Cohen lives in Murray, Kentucky, and Tucson, Arizona.

Interactive Press
The Literature Series

Photograph by Dan Cohen

A PLACE TO READ: LIFE AND BOOKS

MICHAEL COHEN

Interactive Press
The Literature Series

Interactive Press
an imprint of IP (Interactive Publications Pty Ltd)
Treetop Studio • 9 Kuhler Court
Carindale, Queensland, Australia 4152
sales@ipoz.biz
ipoz.biz/IP/IP.htm

First published by IP in 2014
© Michael Cohen, 2014

Printed in 12 pt Book Antiqua on 14 pt Century Gothic.

National Library of Australia Cataloguing-in-Publication entry

Author: Cohen, Michael, author.

Title: A place to read : life and books / Michael Cohen.

ISBN: 9781922120922 (paperback)

Subjects: Cohen, Michael.
                Men--United States--Biography.
                American essays--20th century.
                Books and reading--United States.

Dewey Number: 973.9092

Also by Michael Cohen:

Murder Most Fair: The Appeal of Mystery Fiction (Fairleigh Dickinson University Press/Associated University Presses, 2000).
Sisters: Relation and Rescue in Nineteenth-Century British Novels and Paintings (Fairleigh Dickinson University Press/Associated University Presses, 1995).
Hamlet in My Mind's Eye (University of Georgia Press, 1989).
Engaging English Art: Entering the Work in Two Centuries of English Painting and Poetry (University of Alabama Press, 1987).
The Poem in Question. with Robert E. Bourdette, Jr. (Harcourt Brace Jovanovich, 1983)

Contents

Acknowledgments

I want to thank the editors of the following magazines, in whose pages some of these essays first appeared: *Able Muse, Bicycle Review, Birding, The Bayou Review, The Briar Cliff Review, Harvard Review, The Humanist, The Kenyon Review, The Missouri Review, Monkey Puzzle, The New England Review, New Madrid, North Dakota Quarterly, Southern Humanities Review*, and *34th Parallel*.

The 1952 Jock Carroll photograph of Marilyn Monroe is reproduced by license from CMG Worldwide.

Short quotations appear in the text from the following sources:

Michel de Montaigne, *The Complete Works of Montaigne: Essays, Travel Journals, Letters*. © 1943, 1971, 1976 by Donald M. Frame, © 1948, 1957, 1958 by the Board of Trustees of the Leland Stanford Junior University, published in 1958 by Stanford University Press.

Stuart Hampshire, "Introduction" to *Michel de Montaigne: The Complete Works*, © 2003 by Everyman's Library.

Michel Guillaume de Crèvecoeur writing as J. Hector St. John, *Letters from an American Farmer* (1782).

N. Scott Momaday, *House Made of Dawn*, © 1966, 1967, 1968, 2010 by N. Scott Momaday, published in 1968 by Harper & Row.

Gaston Bachelard, *The Poetics of Space*, © 1958 as *La poétique de l'espace* by Presses Universitaires de France. Translation by Maria Jolas, © 1964 by Orion Press.

Leo Tolstoy, "The Death of Ivan Ilych," translated by Louise and Aylmer Maude (1886).

Marcel Proust, *Remembrance of Things Past*, vol. 1, translated by C.K. Scott Moncrieff & Terence Kilmartin, translation © 1981 by Random House, Inc., published in 1982 by Vintage Books (Penguin Random House).

Ray Sharpe, *Linda Lu*, © 1959 by Ray Sharpe, Peermusic Publishing.

Marty, United Artists 1955, screenplay by Paddy Chayefsky.

Justin Kaplan and Anne Bernays, *The Language of Names: What We Call Ourselves and Why It Matters*, © 1997 by Justin Kaplan and Anne Bernays, published in 1999 by Touchstone (Simon & Schuster).

T. S. Eliot, "The Naming of Cats," in *Old Possum's Book of Practical Cats*, © 1939 by T. S. Eliot, © 1967 by Esme Valerie Eliot, published in 1967 by Harcourt Brace Jovanovich.

Jane Morgan, Christopher O'Neill and Rom Harré, *Nicknames: Their Origins and Social Consequences*, © 1979 by Law Book Company of Australasia.

James Thurber, "The Case Against Women," in *Let Your Mind Alone!* © 1935, 1936, 1937 by James Thurber, published in 1937 by Harper & Brothers.

John Keats, letter to his sister Fanny Keats, March 13, 1819, in *Letters of John Keats to His Family and Friends*, edited by Sidney Colvin, published in 1891 by Macmillan.

Malcolm X and Alex Haley, *The Autobiography of Malcolm X*, © 1964 by Alex Haley and Malcolm X, © 1965 by Alex Haley and Betty Shabazz, published in 1965 by Grove Press.

Barbara Holland, *Endangered Pleasures: In Defense of Naps, Bacon, Martinis, Profanity, and Other Indulgences*, © 1995 by Barbara Holland, published in 1995 by Little, Brown and Company.

Virgil, *The Aeneid: A New Verse Translation by Rolfe Humphries*, © 1951 by Charles Scribner's Sons.

Eavan Boland, "The Latin Lesson". © 2005, 1990 by Eavan Boland, from *New Collected Poems* by Eavan Boland. Used by permission of W. W. Norton & Company, Inc.

Anne Fadiman, "You Are There," "Marrying Libraries," and "Never Do That to a Book," in *Ex Libris: Confessions of a Common Reader*, © 1998 by Anne Fadiman, published in 1998 by Farrar, Straus and Giroux.

Robert Louis Stevenson, *The Silverado Squatters* (1884).

Vladimir Nabokov, *Speak, Memory*, © 1951, 1967 by Vladimir Nabokov, published in 1951 by Victor Gollancz.

Plato, *The Republic*, translated by Benjamin Jowett (1871).

Robert Burton, *The Anatomy of Melancholy* (1621).

"Norwegian Doctor Takes Patients Who Only Need Understanding," *Wall Street Journal*, January 19, 1996.

Elizabeth Gaskell, *Cranford* (1853).

Nancy MacDonell Smith, *The Classic Ten: The True Story of the Little Black Dress and Nine Other Fashion Favorites*, © 2003 by Nancy MacDonell Smith, published in 2003 by Penguin Books.

Anne Hollander, *Sex and Suits: The Evolution of Modern Dress*, © 1994 by Anne Hollander, published in 1994 by Alfred A. Knopf.

H. L. Mencken, "Capital Punishment," in *Prejudices: Fifth Series*, © 1926 by Alfred A. Knopf.

Scott Turow, "To Kill or Not to Kill," *The New Yorker*, January 6, 2003.

Gary Wills, "The Dramaturgy of Death," *The New York Review of Books*, June 21, 2001.

Aeschylus, *The Oresteia*, translated by Robert Fagles, © 1966, 1967, 1975 by Robert Fagles, published in 1975 by Viking Penguin.

La Bamba, traditional Mexican folk song of uncertain date.

Alain de Botton, *The Art of Travel*, © 2002 by Alain de Botton, published in 2002 by Random House.

Susan Sontag, *On Photography*, © 1977 by Susan Sontag, published in 1978 by Farrar, Straus and Giroux.

Walt Whitman, "This Compost," in *Leaves of Grass* (1867 edition).

Stephen Greenblatt, *Shakespearean Negotiations*, © 1988 by the regents of the University of California, published in 1988 by the University of California Press.

Charles William Eliot, "Introduction," to the *Harvard Classics* (1909).

Robert Hutchins, speech at the launch of Encyclopedia Britannica's *Great Books of the Western World*, April 15, 1952. See Milton Meyer, *Robert Maynard Hutchins: A Memoir*, © 1993 by the regents of the University of California, published in 1993 by the University of California Press.

Denis Diderot, and Jean le Rond d'Alembert, *Encyclopédie, ou dictionnaire raisonné des sciences, des arts et des métiers* (1751-1765).

Adam Kirsch, "The Five-Foot Shelf Reconsidered," in *Harvard Magazine*, 2001.

David Sedaris, "Memento Mori," *The New Yorker*, May 8, 2006.

Samuel Clemens as Mark Twain, *The Innocents Abroad, or The New Pilgrims' Progress* (1869).

Joseph Conrad, letter to J. B. Pinker, February 21, 1906, *The Collected Letters of Joseph Conrad*, vol. 3, © 1988 the Estate of Joseph Conrad, published in 1988 by Cambridge University Press.

Anne Frank, entry for November 11, 1943, *The Diary of a Young Girl: The Definitive Edition*, edited by Otto H. Frank and Mirjam Pressler, © 1991, 2001 by The Anne Frank-Fonds. English translation by Susan Massotty, © 1995, 2001 by Doubleday, a division of Random House.

John Windsor, "Fountain Pen Converts Are Doing the Write Thing," in *The Observer*, July 21, 2007.

Samuel Hynes, "The Feeling of Flying," in *The Sewanee Review*,

Spring, 1987.

Willa Cather, *Death Comes for the Archbishop* (1927).

John Milton, *Areopagitica* (1644).

Washington Irving, *The Sketch Book* (1820).

Walter Benjamin, "Unpacking My Library," in *Illuminations*, © 1955 by Suhrkamp Verlag, Frankfurt, English translation by Harry Zohn © 1968 by Harcourt Brace Jovanovich, published in 1968 by Harcourt Brace Jovanovich.

A. R. Ammons, "Corson's Inlet," in *Collected Poems*, © 1972 by A. R. Ammons, published in 1972 by W. W. Norton.

Sylvia Grider, "Should Roadside Memorials Be Banned?" in *New York Times*, July 12, 2009.

Virginia Woolf, excerpt from *The Letters of Virginia Woolf*, Vol. II: 1912-1922, © 1976 by Quentin Bell and Angelica Garnett. Used by permission of Houghton Mifflin Harcourt Publishing Company. All rights reserved. Excerpt from *The Diary of Virginia Woolf*, Vol. III: 1925-1930, © 1980 by Quentin Bell and Angelica Garnett. Used by permission of Houghton Mifflin Harcourt Publishing Company. All rights reserved.

Lewis Thomas, "Why Montaigne Is Not a Bore," in *The Medusa and the Snail: More Notes of a Biology Watcher*, © 1974, 1975, 1976, 1977, 1978, 1979 by Lewis Thomas, published in 1979 by The Viking Press.

E. B. White, *Charlotte's Web*, © 1952 by E. B. White, published in 1952 by Harper and Brothers; *Essays of E. B. White*, © 1934, 1939, 1941, 1947, 1949, 1954-1958, 1960, 1961, 1963, 1966, 1971, 1975, 1977 by E. B. White, HarperCollins Publishers; *One Man's Meat: A New and Enlarged Edition*, © 1938-1944 by E. B. White, published in 1944 by Harper and Brothers; *The Points of My Compass: Letters from the East, the West, the North, the South*, © 1954-1958, 1960-1962 by E. B. White, HarperCollins Publishers.

Roger Angell, *Let Me Finish*, © 2006 by Roger Angell, published in 2006 by Harvest Books, Harcourt, Inc.

Henry James, *The Portrait of a Lady* (1881).

Henry Beston, *The Outermost House: A Year of Life on The Great Beach of Cape Cod*, © 1928, 1949 by Henry Beston, published in 1949 by Holt Rinehart.

Tom Lewis, *Divided Highways: Building the Interstate Highways, Transforming American Life*, © 1997 by Tom Lewis, published in 1997 by Viking Penguin.

Preface

I used to write academic books. When I retired from teaching, I began to write and publish personal essays. Some of the published essays, plus a few that have not seen print before, make up the chapters of this book. It is an accidental memoir composed of looks backward at my life from the perspective of my seventh decade, not a chronological record of what happened to me. Many of these chapters have more to say about my reading than about my life, although my point is that this may often be a distinction without a difference.

Why the essay form? I think, for me, the reason is that essays tell a kind of truth not found elsewhere. It's not quite the literal truth many people mistakenly equate with all nonfiction writing, but a truth that only emerges in the act of writing and in this form.

Scott Russell Sanders starts his essay "The Inheritance of Tools" with a coincidence: at the moment his father is dying, the author is using one of the tools his father gave him, and in fact it is causing him pain: he hits his thumb with a hammer. "Did this really happen?" the cynical reader may be forgiven for asking. Of course it does not matter whether it happened or not. An essay is neither biography nor journalism. Even if it were the one or the other, the experienced reader expects some invention, imagination, or adjustment of the "facts". The writer's page is the artist's blank sheet. "Nobody knows what's 'really there' until you draw it", my old drawing teacher used to say. When a student defended a confusing or ambiguous shape by saying "But that's what was there!" "You're responsible for what's on the page you create," he'd say, "not some accident of nature."

Essays offer a kind of truth available nowhere else, a truth that comes in the juxtaposition of experiences, placed by the mind in different arrangements until connections become clear. Essay writing is for me a mode of discovery that facilitates and enlarges memory; I find out more about myself in the writing than I knew before. So do my friends and family. "Amusing notion," writes Montaigne, "many things that I would not want to tell anyone, I tell the public; and for my most secret knowledge and thoughts I send my most faithful friends to a bookseller's shop."

Dan and Matt and Katharine Cohen have been extraordinarily helpful during the writing of this book and to them I give my thanks and love. Thanks also to David Earnest, who read every word and caught many errors; all those that remain are mine.

1. Location, Location, Location

In the middle of my new parlour I have, you may remember,
a curious republic of industrious hornets... My family are so
accustomed to their strong buzzing, that no one takes any notice
of them.

– de Crèvecoeur, *Letters from an American Farmer* (1782)

Sometimes, waking suddenly, I don't know where I
am. The room, the city, even the country will not come
clear; they wait for the conscious mind's wheels to take
hold on the firm ground of the here, of location. Waking
slowly, I never have this limbo moment of unlocation.
While my mind rises from unconsciousness to consciousness, vibrations, the air's wetness or dryness, the
room's closeness or expanse, and above all, the sounds
have located me firmly before I reach the surface. Every location has these distinctive markers. Even in the
"quiet" of night there are sounds that say unequivocally, "This is a west Kentucky June midnight", or "It is
just before dawn in the Tucson Mountains".

For these are the two places where I spend most of
my time. Not that I couldn't be waking elsewhere —
some town where I lived months of my life, such as
Ankara or New Orleans, London, Madrid, or even
Springfield, Ohio, where my family spent a miserable
but memorable winter when I was seven. I could even
be waking in a place I've *never* been before. One has
those wakings, caused by the slowing of the train and
the change in noises; peering out of the always grimy
windows with eyes blurred by sleep and lack of eye-
glasses (Where are they? Good Lord, tell me they're not
in the berth with me and that I haven't crushed them!),

1

one sees the spire of a church sliding leftwards in the middle distance. Is this York? Is that the dome of St. Eustace? Can we have reached Edinburgh already? How far have I come? As much as I enjoy it, traveling is a transition, and I am not myself while traveling. But the usual mechanism works to try to locate me. It is as if clear perception and thinking were tied to knowing where one is, like those telescopes that start with an extremely accurate fix of their locations derived from the satellites of the Global Positioning System. They listen for signals, triangulate, find exactly where they are, and then they can "think", can find any star or other celestial object in their memories.

These two places where I live: the Blood River basin on Kentucky Lake and the eastward slope of the Tucson Mountains, though they are so different, have shaped my experience and perception. They also have made notches — default settings, if you will — in whatever interior position finder I have. I grew up in Tucson, but more importantly, spent my college years there, married there, and my first son was born there. A landscape where the eye can reach unobstructed to the horizon always seems more comfortable to me than one where trees or hills stop the eye. But for thirty years, almost all of my working life, I have lived in Kentucky, where I learned about deciduous forest, wide river bottoms and lakes, unsubtle seasonal change. Sailing became a new love, and clear winter skies far away from any city enchanted me. In both these places I feel located. I suspect some of the new regionalists like Wendell Berry, reading that last sentence, would snort that it is not possible to be located in two places, that such rootlessness is part of the reason the country is in such a pitiful condition. My own sense of being located differs from theirs in an important way: the ties I have to place were not bound up before I was born, nor yet in early childhood. I built attachments to these places through an adult's reflection and association. And location for me is a present state. But this is to simplify too radically the very complicated notion of where I am "at home" and how I locate myself. Locating turns out to be a process that is not independent of time and culture, although for me, it is remarkably free of the usual associations of home.

Let us begin by being precise. My house on the Blood River is at north latitude 36° 02′ 24″ and west longitude 88° 04′ 41″. My house in the desert is at north latitude 32° 16′ 30″ and west longitude 111° 03′ 38″. I know these numbers because I have studied a navigation chart of Kentucky Lake (mile 41.5 to mile 57.8) and a topographical map of the Tucson Mountain area that includes my house there. Why? Because I am interested in what can be seen in the sky. Because, as it turns out, internet programs that will tell you what's in the night sky at any given time — right down to the exact derivation of the space junk you may see orbiting overhead (is it a Russian Okean satellite or a spent booster rocket?) — these programs ask for your exact location. Your location determines the precise time and the precise direction in the sky where you will see transient phenomena like orbiting space junk or comets as well as perennially recurring sights such as the constellation of Hercules at its zenith.

Latitude and longitude constitute a simple system of coordinates for locating any point on the earth, starting from an imaginary circle around the earth's slightly bulging middle — the equator — and a circle that goes through the poles and Greenwich, England. This system is at once abstract (composed as it is of arbitrary and imaginary lines) and mortally specific: satellites in stationary orbits beam signals to receivers on Earth that can direct a bomb to within a few yards of any dictator's desk or child's cradle. Such accurate mayhem was not the purpose of latitude and longitude, of course. The coordinate system was intended for travelers; it is an attempt to locate us, in transit, on a global scale. This is, of course, to simplify and to make innocent what has never really been so. Maps are the tools of states, produced by kings and used by them to appropriate. The Global Positioning System, which enables us to find our location within feet anywhere on earth, was developed to direct weapons, and until just a few years ago, all civilian GPS devices had errors deliberately built into them because the Pentagon feared they would be used by enemies with intentions just as bellicose as those of the inventors. And longitude was not a useful coordinate until the British government bribed inventors to come up with an accurate timepiece so that it could be measured, and measured mainly to preserve British superiority on the high seas.

Terrestrial and celestial ways of locating are connected. For example, the easiest way to find your latitude quickly in the northern hemisphere is to measure the angle between the horizon and Polaris, the North Star. That angle *is* your latitude. But in order to find longitude, measuring the height of stars is not enough; one must also know what time it is. Longitude itself can be seen as a measure of time. Imagine the earth as the center of a giant clock dial, and imagine a single clock hand projecting up into the sky from Greenwich, England — where the longitude is zero degrees. In twenty-four hours that clock hand will sweep the whole sky and be pointing again to the same place it started. That clock hand travels, in other words, at just half the rate of an ordinary clock's hour hand. At Greenwich, all we would have to do is look at the stars' positions to know what time it is. But if we were not at Greenwich, we would have to know the stars' positions *and the time* to know how far west or east of Greenwich we were — that is, our longitude. In a universe where everything is in motion, location is dependent on time. These days, of course, Greenwich is nowhere near where the political measurements start, but at one time it was, and that's why the longitude measurement started there as well.

On a more human scale, the Kentucky house is situated about thirty feet above Kentucky Lake, more specifically a bay on the lake named for the Blood River that feeds into it. Our rise has a parallel ridge across the water, green at its peak even in winter. Though it's called Kentucky Lake, this long and narrow impoundment basin was created by damming the north-flowing Tennessee River in 1943. During my first summer in this house thirty years ago, I beached my little daysailer on the limestone scrabble of the far shore of the Blood River and climbed the ridge there, through the hickory and oak trees into the pine grove on the crest. There I found a tiny cemetery with two dozen graves, their markers all of punched tin and all dated 1943. My first reaction was that it must have been the resting place for a local military unit wiped out somewhere in the Pacific in the middle of World War II; then I realized that I was in a removal cemetery of graves that had stood in the flood plain of the new lake, near the then small tributary of the Tennessee called the Blood River, so called because the ubiquitous red clay, not the blood of its young men, stained the waters.

The Blood River house is the way I think of it, BR for short. Coincidentally (is it Freud who says there is no such thing as coincidence?) these initials are also our way of referring to the Brown Recluse Spider, our perennial roommate in the Blood River house, there the day we moved in and never eliminated completely by the spraying or fumigation of the house that we do at intervals of five or ten years. We capture the spiders in a glass and take them outside when we find them in closets or bathrooms—generally two or three a week. Though their bites can be dangerous, sometimes causing a necrotic destruction of much of the flesh around the wound, ours have never, so far as I know, bitten anyone in the house.

At one point my wife Katharine and I considered officially naming the house. The president's lodging at my little college is pretentiously called Oakhurst, so we thought Pighurst might do for our house. Another possible name was Camp Cohen. When we were younger and entertained the friends with whom we'd taught in New Orleans, sometimes five families at once, everyone called the place Camp Cohen. But the Blood River house is the way I think of it.

The desert house also has its tentative name and its creatures. Many years ago, someone put up a sign that read "Guard Dog" to deter burglars and weekend fun-seekers who cruise dirt roads in the desert. The house is isolated, or was; other houses have encroached and now there are newcomers within a hundred yards where before it was half a mile to the closest house. Some time after the first sign was put up, a friend commissioned a tile that reads "Guard Dog Ranch". It is not a ranch, though there is a corral, a well, and a few acres of mesquite and chaparral. Katharine likes to refer to the place as GDR. I prefer to call it just the desert house.

As for creatures, there are coyotes and javelina, neither dangerous to humans. One night a Western Diamondback, a rattlesnake gave its unmistakable warning, like a tiny maraca, to my son, who had opened the patio door to step outside; snakes occasionally get into the walled back yard. The house's smaller denizens are potentially more dangerous than the snakes or the bobcat that sometimes gets up on the roof: we share the indoor space with a few scorpions, sun spiders, and the ubiquitous kissing bugs. Some readers with monthly pest control and airtight houses will doubtless be dismayed by my account of these poisonous creatures or by the seeming nonchalance with which we live among them. But we are careful. We watch where we put our hands and our behinds (scorpions love toilets — water being always at a premium for desert creatures). And much must be said for being used to it. Melville has a wonderful chapter in *Moby Dick* called "The Line", which describes the long rope attached to the whaler's harpoon. The line snakes in and out among the rowers in the whaleboat and may at any moment, if the harpooner hits a whale, begin to run out at incredible speed, posing obvious dangers to all the crew members near it, and every crew member in the tiny whaleboat is near it. "If you be a philosopher," writes Melville, "you will realize that you always face risks, even sitting at home in seeming safety." His point is that most of us don't think we're at risk at home, though we would know we were at risk in that whaleboat. My little creatures help remind me I'm at risk at home, just as you and I are at risk anywhere.

I am located when I am in one of these places. I know that in Tucson, the North Star is over the small Palo Verde tree behind the back wall of the patio; on the Blood River, the North Star is nearly in line with the lake-facing side of the house. I know where the sun rises in the summer and where in the winter, because, as I said, in such a universe as this, moving "in all of the directions it can whiz" (as "The Galaxy Song" has it), location is tied up with time.

The connection of space — locations where one lives — and time has fascinated writers of fiction as well as philosophers. In N. Scott Momaday's *House Made of Dawn* (1969), a grandfather takes his grandsons out into the first light of dawn and faces them toward the black mesa to the east. He tells them they must learn its shape and where along this shape the sun rises at each part of the year, planting and harvest, hunts and celebration times. "They must know the long journey of the sun on the black mesa, how it rode in the seasons and the years, and they must live according to the sun appearing, for only then could they reckon where they were, where all things were, in time..." The temporal quality of location cannot be neglected, nor yet the transience of the place and of the person in it. The philosopher Gaston Bachelard theorizes about one's house and its significance in *The Poetics of Space* (1958). The house in the imagination, Bachelard says, is where our memories are housed; it is a "simple localization of our memories." "At times," he continues, "we think we know ourselves in time, when all we know is a sequence of fixations in the spaces of the being's stability... In its countless alveoli space contains compressed time. That is what space is for." This idea of space as a receptacle of time and memory is the theme, also, of Proust's *In Search of Lost Time* (1913-1927), where places (Combray, Balbec, Paris) or paths through places ("Swann's Way" or "The Guermantes Way") are the containers of memory, which can be evoked by revisiting the place or traveling along the path. Bachelard specifically points to the ancient memory aid of associating a series of things we want to remember with rooms in a house we know well. Though he does not mention Proust, *In Search of Lost Time* monumentally illustrates this way memory works.

As living spaces my places were produced historically by appropriation—we could probably say the same (certainly Nietzsche would) of all living spaces. The Blood River house sits on a ridge of rocky clay soil that would never have been cleared of its deciduous oak and hickory forest (with a few pines mixed in) for houses or even for farmland. Even bottomland is barely farmable in this part of the county. But the Blood River lies only a few hundred yards to the east, and the Blood River runs into the Tennessee — which the Army Corps of Engineers and the Tennessee Valley Authority decided to dam after the great flood of 1937. They built a dam twenty-five miles to the north and filled the whole basin of the Tennessee and its tributaries with water, evicting hundreds from bottomland farms and small settlements. Eventually, when Kentucky Lake's twin, Lake Barkley, was created by damming the Cumberland, whole towns were displaced. As a result of these events and others they put in motion (the growth of the little teachers' college that became my university, for instance), I now have a waterfront house on the Blood River, no longer a river but a half-mile wide bay of Kentucky Lake.

In Arizona, the Santa Cruz River Valley was for more than a millennium the home of a people now known only as the Hohokam, a name given them by the Pima Indians who replaced them. Hohokam means "those who are gone", and indeed, there are scarcely more than a few pottery fragments to mark the places where Hohokam settlements dotted the valley and nearby uplands until about 1500 A.D., when the Hohokam mysteriously disappeared. The later settlers, the Pima and the Tohono O'odham, ranged freely across the river valley and the mountain ranges of this part of Arizona until they were "put in their place", first by the Spanish missionaries, starting with Eusebio Kino, then by Mexicans after they won their independence from Spain, then by the U. S. Army, whose Tucson encampment, Fort Lowell, was built over the remains of a Hohokam extended village established more than fifteen hundred years ago—a few miles east of where my house now stands. The Tucson city founders got no resistance from the by-that-time pious and peaceful Indians who came after the Hohokam and, who after all, had never been raiders like their Apache cousins. Then, using a map like a sword, the U. S. Government restricted the Indians to their reserved lands, in this case, on the other side of the Tucson Mountains from where my house now stands.

But the physical and historical spaces of my houses are not their only locators; each has also a dimension of mind. Milton's Satan, speciously arguing that hell can be made into heaven, is at least partly right that "the mind is its own place." Conversely, places are at least partly mental spaces. And here I believe myself peculiarly at odds with the usual mental associations and connotations of places we call home. For the theorist of space, Gaston Bachelard, all inhabited places, even the merest shelters, partake of the features imagination ties to the concept of "home": in dreams and memory and in the images of poetry, the house is maternal, comforting; it has "the illusion of protection". The feminine and protective ideas attached to "home" are romanticized in their most famous nineteenth-century statement by John Ruskin.

In an 1864 lecture called "Of Queens' Gardens", Ruskin gives the sentimental view of home as ultimate refuge and sanctuary:

> This is the true nature of home—it is the place of Peace, the shelter, not only from all injury, but from all terror, doubt, and division. In so far as it is not this, it is not home: so far as these anxieties of the outer life penetrate into it, and the inconsistently-minded, unknown, unloved, or hostile society of the outer world is allowed by either husband or wife to cross the threshold, it ceases to be home; it is then only a part of that outer world which you have roofed over, and lighted fire in. But so far as it is a sacred place, a vestal

temple, a temple of the hearth watched over by Household
Gods, before whose faces none may come but those whom
they can receive with love, — so far as it is this, and roof
and fire are types only of a nobler shade and light, — shade
as of the rock in a weary land, and light as of the Pharos in
the stormy sea, — so far it vindicates the name, and fulfills
the praise, of home.

In Ruskin's view of home, the impossible shining
shield that surrounds it and the impossibly idealized
woman within are synonymous; she somehow generates the shield. We are a little less patient with this pa-
tronizing notion about women as sacred homemakers
now, but Ruskin's words about the sacred temple of
the home still resonate with many people. I cannot go
along with either of his notions — that where one lives
is profoundly distinct from the outside world, far less
that it should be an extension of one's person. Those
who see their houses as personal extensions try to create a structure around themselves that is completely
those selves: each householder generates his own shining shield. The house becomes a material representation of the owner, an exudate of his personality. Rather
than being *in a place* when we go home, we enter into
ourselves. Home thus becomes a material extension
of the modern cult of individuality. The notion seems
to be first of all meretricious: what we generate, I believe, is more likely to be a reflection of consumer culture than of some original essential being of our own.
Tolstoy, in *The Death of Ivan Ilych*, pitilessly describes
Ilych's "home improvement", and his complacent belief that he has made his house distinctive. "In reality
it was just what is usually seen in the houses of people
of moderate means who want to appear rich," writes
Tolstoy. "His house was so like the others that it would
never have been noticed, but to him it all seemed to
be quite exceptional." Some may indeed successfully

11

create in a house an image of a self, an image that is at once original and distinct from advertising culture. But here, what happens is the substitution of an awareness of oneself for an awareness of place.

As for myself, I fear that I have indeed, in my houses, merely "only a part of that outer world" which I have "roofed over, and lighted fire in." But that outer world is what I seek to know, rather than merely to mount a mirror in place of it, and build a wall around the mirror.

The Blood River house was built by a Mr. Moore, whose son had just started to teach at Murray State University. For many years after we first moved in — we bought it the year after I began teaching at Murray State — people in the neighborhood still referred to it as "the Moore place". The desert house belonged to Katharine's parents before it was ours, but it was built by yet another family. I don't think of myself as a squatter, exactly, but I do feel my own presence as more temporary than, if not these houses, then certainly the oaks that surround the Blood River house. And I have not changed the houses much. "Late in life," writes Bachelard, "with indomitable courage, we continue to say that we are going to do what we have not yet done: we are going to build a house." Everyone, according to Bachelard, has a "dream house", by which he means a house composed of memories, an imaginative ideal of the past, but he also means what we ordinarily mean by this cliché: a perfect house that we would like to build, which contains all those things we have liked in other houses. Everyone is commonly imagined to have such a dream house. But I don't. What does it mean not to have a dream house? I think it means that I do not seek in a house what is commonly associated with the notion in dream, memory, popular imagination: the "illusion of protection", as Bachelard so aptly puts it, and the warmth and coziness of the space of the mother. I am strangely content with houses built by others. A house is for me, like existence itself, where I find myself, where I have ended up. Of course I modify the space; merely being in it modifies it. But my changes are fewer than my tolerances for what is. Thoreau, not one of my favorite thinkers, nevertheless expresses my thought exactly when he calls a man rich in proportion to what he can afford to leave alone. And I think I am also of Montaigne's mind, when, in "Of Vanity" he writes about being content with his father's estate at Montaigne, completing a wall his father started or repairing some building, not really ablaze to put his own mark on the place.

Knowing where I am is just that; location isn't refuge or self-expression. Nor is it safety or permanence, though these are qualities we may unthinkingly ascribe to being at home. Neither quality inheres in the buildings of the Blood River house or the desert house, nor yet even in the trees or the stones. When we moved into the Blood River house, a river birch tree stood thirty feet or so from the water. This tree was a locating mark when we approached by boat, and in spring we often picnicked under its branches. It seemed sturdier, somehow, than the oaks farther up the water's edge, built for stress and change. When high winds took its top branches out, it looked distressed for a season; then it rallied and swelled out in a new, squatter shape. Last year the water, driven by the prevailing north wind, began to undercut the birch tree's roots from the bank. This year the tree is gone.

Behind the desert house, a small cairn of stones, perhaps ten inches high, marks a bend in one of the horse trails. My father-in-law, who lived in this house before me, piled these stones, picking up a suitably-sized rock every few days as he approached the cairn from either

13

direction along the trail, piling them loosely but purposefully. Then one day, perhaps the very day he had the small stroke that was the beginning of his final illness, he placed the last stone there. Horses occasionally kick a stone away, and eventually the spot will look exactly like the scrabble around it—unless, of course, I take to piling up rocks there myself. The houses themselves seem to have more permanence, but in the desert's or the river's scheme of time they are no different from the cairn of stones. Proust writes at the end of *Swann's Way* that "houses, roads, avenues are as fugitive, alas, as the years." His narrator reminds us of this impermanence of places and how we try to save them in "the little world of space on which we map them for our own convenience", whether our maps are those of memory, of a novel, or of an essay.

Well before daylight the first Mourning Dove begins to call. First come two notes, then its whole mournful yodel. For a while only one dove sings. Later answering calls come. There are pauses, as would not be the case in full spring. After some minutes the single note of a Gambel's Quail joins the chorus. Now the eastern sky has barely lightened. Finally, as some light begins to touch the western part of the sky's dome, the more enthusiastic song of the house finch joins the other two voices.

Though I am still not fully awake, these impressions have registered on my stirring consciousness. This is not the noisy, multi-layered dawn chorus of a Kentucky morning at any time, nor yet the sparer but still exuberant April or May morning song of Arizona birds. It is March, just before dawn in the Tucson Mountains. I know where I am.

2. Names

Well now they call my baby Patty but her real name her real
name her real name is Linda Lu.

— *Linda Lu* (1959)

The names in my mother's family are difficult to explain to outsiders. My mother is called Dot, although
her name is not Dorothy but Agnes — Agnes Emelda, to
be exact. But she has always been Dot to her family. Dot
is conventionally a nickname for Dorothy; my mother
has a sister whose name is Dorothy, but no one has ever
called her Dorothy or Dot or Dottie; her name in the
family has always been Dick. My son was trying to explain this to one of his college chums one day, and the
chum replied with great glee, "You've got a great-aunt
Dick. Wow! *I've* got a great uncle Hyman!"

I.

I hated my nickname, "Marty", when I was growing up. My middle name is Martin, and because one
of my mother's sisters in a nearby town named her
newborn "Michael" six months before I was born, my
mother decided not to call me by my first name, but to
use my middle name instead. The confusion of having
two Michaels was probably overrated; at least I don't
remember ever having much to do with my cousin and
his family, and we moved away when I was six. But the
damage was done; Martin was shortened to Marty, and
Marty I became.

I have never thought nicknames ending in –y were
anything other than demeaning; it is, after all, one of

the ways we make a word into a diminutive. There are other reasons for not burdening a child with an -ie or -y name. If your daughter is called "Kitty," do you think there is any chance her schoolmates will NOT call her "Kitty Litter"? Southerners regularly name their children "Jody" (the other kids will yodel it) or "Hughie" (instantly transformed into "Huge-y" to the chagrin of the child, pudgy or not). Of course, one "Jimmy" (a name rather than a nickname, in this case) became president, but this one success cannot disprove any rule about belittling names, since Carter's voters elected him despite it, rather than because of it.

Paddy Chayefsky's play *Marty*, made into a movie starring Ernest Borgnine in 1955, finally decided me to reclaim my real name. The name of Borgnine's character, who was already verging on middle age, was used incessantly: "What do you wanna do tonight, Marty?" The emptiness of the character's life seemed inexorably linked to his neotenous name. When I entered high school I began introducing myself as Michael or Mike, and always using Michael on written work. Not everyone cooperated with my efforts; friends of the family persisted in calling me Marty, and my friends from grade school thought it was "weird" that I was insisting on my real first name. Pubescent and post-pubescent children are often sensitive about their names, an aspect of their extreme self-consciousness. Parts of them are suddenly changing, and they seem to have even less control than usual of their selves, their identities. At such a time, insisting on one's name instead of a nickname may be an attempt to assert some control. Looking back, I think this wish for control was part of what I did, but I have to think it was partly an aesthetic matter, too.

One of my contemporaries decided to change her name completely when she was in high school. She got tired of her typically Jewish given name and surname, announcing to all her friends and teachers that henceforth she would be known as Cynthia Wentworth. They were astonished, amused, scornful, but she persisted and bullied everyone until they gave in out of weariness. Her name *became* Cynthia Wentworth. Her classmates and mine thought changing one's name was something only celebrities did. If your name is Arlington Spangler Brugh (my mother-in-law's schoolmate from Beatrice, Nebraska) and you don't think Hollywood will like it, you change it to Robert Taylor, and no one is surprised at the simplification. Or if your name is *too* simple, say "Prince", you can complicate the issue and become "The Artist Formerly Known as Prince".

In my case, when I went away to college, none of my new friends knew me as anything other than Michael, and Marty became a name no one except my siblings and my mother used for me. But then a few years later, when I was in graduate school, I had a strange experience: I answered a telephone call and the voice at the other end said "Hello, Mahty?" It was a local call (in those days long-distance calls *sounded* as if they were coming from a long distance) and no one in that town knew me as Marty. I hesitated. "Is this the Mahty Cohen who woiks in cycles?" asked the caller. This threw me even more. How could this strange-accented caller know about my work habits? As it turned out, there was a Marty Cohen who worked in a motorcycle shop downtown, and it was this Marty the caller sought. But the call gave me the feeling I could not escape my early nickname so easily.

II.

According to Justin Kaplan and Anne Bernays' 1997 book, *The Language of Names*, I was exercising the

17

American right to rename and refashion myself. But I did not think of my insistence on my real name as "renaming": I was reclaiming my own name. These authors say the prime example of American name reclamation is that of the Indians. Kaplan and Bernays write that the European settlers' persistence in calling the natives of America "Indians" is "one of the great naming blunders of history", and that the blunder persists to the extent that the movement to insist on "Native Americans" has almost disappeared, "even in tribal councils." But this blunder illustrates the general problem of naming, which is so frequently done by others, from the outside. American Indians are not peoples of subcontinental India, of course. But neither are they "native Americans" unless that phrase means merely being born in America, in which case they do not distinguish themselves from me: I am a native American. American Indians are what they are—what they named themselves: Lakota, Tohono O'odham and so on. But are they not also American Indians? I can also be located by names not necessarily chosen by the immigrant peoples to whom they were attached: Jewish, Irish (since my mother was a Schlairet). I always think of myself, though, like Salinger's Seymour Glass, as a Jewish-Irish Mohican Scout.

There is a vanished tribe called the Hohokam, a group of Indians that preceded the Pima and the Papago in the middle of Arizona. Not far from where I now live, they left pots and other artifacts in a settlement from which they disappeared seven hundred years ago. The Hohokam, which literally means something like "the people who are gone", were named by the Pimas; we have no idea what they called themselves. If you are not around to protest, you get the name others give you (a constraint also true of those too young to protest the names and nicknames they're given). The

Papago in recent decades have renamed themselves
Tohono O'odham, which means "the desert people".
They profoundly disliked the name Papago, also given
them by the Pima, and meaning "bean-eaters".

The issue of reclaiming or renaming, Kaplan and
Bernays characterize as a conflict in attitudes between
America and the Old World. I agree that it has to do with
issues of self-definition, individualism, and democracy, but these issues cannot claim to have been invented
by Americans. Rumblings against the old order often
show up in satire against aristocratic *names*, because
names embody, solidify, and perpetuate class distinctions and aristocracies. In *Candide*, Voltaire ridicules
the older and younger Baron Thunder-ten-Tronkh's
pride and defensiveness about the seventy-two quar-
terings in their coat of arms; near the story's end the
young Baron is unwilling to allow Candide to marry
his sister even though both are penniless, she is old and
ugly, and the rich Candide is still willing to marry her.
Henry Fielding's narrator in *Joseph Andrews* remarks
that the young foundling Joseph had just as many an-
cestors as the people who scorned him. Robert Brown-
ing's *Duke of Ferrara* huffs that his last duchess acted
"as if she ranked / My gift of a nine-hundred-years-old
name / With anybody's gift."

III.

But most of us now live in a world where money,
rather than names, creates the caste distinctions. What
then, is wrong with the old names, and why would anyone want a new one? Names, first and last, aristocratic and plebeian, are important, and their importance
comes, I believe, from a deep-seated conviction that a
name, if it's genuine, will indicate the true nature, the
essence, of its owner. But only if it's genuine. There are

books with titles like *The Hidden Truth of Your Name: A Complete Guide to First Names and What They Say about the Real You* and *The Secret Meaning of Names.* But we know the books cannot help us to divine what the person is really like, and therefore the name, whatever it means, is as likely to be misapplied as not. We don't know what our real names should be. With our own names we are in the same doubt as the farmer who, when asked the name of his dog, replied, "We don't know, but we call him Spot." The dog doesn't know his real name either. T. S. Eliot insisted in "The Naming of Cats", that the cats do know, and that when you see a cat looking meditative it's because he's thinking about "His ineffable effable / Effanineffable / Deep and inscrutable singular name." But cats don't know either. The real problem with names is that while we hold them in such importance, believe they somehow stand for us, consider them, as Bernays and Kaplan say, "sacred", yet we can't ever get at what they might be in absolute or platonic terms; we can't ever read the ideal list of dramatis personae for the play we're in. We have to be content with either the names we are given – and I was certainly not content with mine – or we have to try with great labor to make or remake our names in an effort of self-fashioning. T. S. Eliot, by the way, was one of many American writers who was less than content with his name: T. S. somehow sounded better to the young poet than Thomas Stearns. Ralph Waldo Emerson added the Waldo while in college, David Henry Thoreau decided Henry David was a better sequence, and of course, Samuel Clemens became Mark Twain.

Complementing the worry that our names are not right for us is the worry that the parent-chosen first names that we dislike, or the schoolmate-chosen nicknames that we detest really *do* capture our essential nature. We have daily assurances that names have power—if not to point to our essential nature, then to embarrass. Flaubert has a classic scene of humiliation the day the young Charles Bovary goes to a new school; he mumbles his name when called on and the students begin to chant what they have heard, making of it a humiliating singsong taunt: "Charbovari! Charbovari!"

We know that certain names evoke reactions or expectations: Marilyn obviously suggested more glamour than Norma Jeane even before there was a star of that name. Our names do go before us. Names are liked and disliked, while the people bearing them are prejudged on an arbitrary, though not completely unpredictable basis. The name "Wendy", for example, according to researchers Jane Morgan, Christopher O'Neill, and Rom Harré, is "disliked by women but liked by men." This trio published a book in 1979 called *Nicknames: Their Origins and Social Consequences*; the seventies not only produced people who changed their names to Preserve Our Mother Earth, but also sociologists who studied names. Morgan, O'Neill, and Harré looked at the way children find nicknames for each other, and the results of their study confirm all those bad memories you have about the way the kids around you went straight for your most vulnerable spot when they gave you a nickname that highlighted your racial difference, the habit that you hoped no one would notice, or your least attractive physical feature. The three researchers also asked subjects to tell them reactions to common names and expectations about what the people bearing these names would be like. Someone bearing the name "Jane" (the name of one of the researchers) was expected by their subjects, stereotypically, to be "plain". But "Christopher" (also the name of one of the three) was expected to be "fair-haired, tall, and charming". The third researcher, Rom, has a name that escaped scrutiny in the study, probably because it is not common.

(There was no agreement among the test subjects about what to expect from a "Michael".)

Parents clearly have to be alert to avoid unpleasant combinations of names. Even Socrates, says Montaigne in "Of Names", considers it "worthy of a father's care to give his children attractive names." My brother was only half joking when he worried aloud about naming one of his girls "Jessica", for fear those to whom she introduced herself as "Jessica Cohen" would think she was stuttering. And my stepfather told the story of overhearing a girl to whom he'd just introduced himself reporting to a girl-friend that "he's cute, but it's too bad he lisps." My stepfather's name was Seth Douthett. But parents, aside from avoiding unfortunate names, can do little more than guess whether the name they choose will actually fit the human being who is before them. They cannot steer the child's future with a name, although the prospect is tempting. My wife and I considered naming our first son Sir Charles. Why wait a lifetime for a knighthood? But of course no one would be fooled. "What prevents my groom from calling himself Pompey the Great?" asks Montaigne.

IV.

Sometimes our desire for control in the matter of our own names is as small as the gesture of changing one letter. My sister decided — I'm not sure exactly when — that her name was going to be spelled "Judyth" instead of "Judith". OK. When I dedicated a book to her a few years ago (appropriately named *Sisters*), I spelled the name her way. But I have difficulty explaining to myself why this adjustment seems more acceptable than the complete change, say, to Kirk Douglas from Issur Danielovich. Your name is your own, but not exclusively your own; it also belongs to others as public record.

My older son is named not Sir Charles but Matthew, and he has no middle name. When my wife Katharine and I came to name our second son, we gave him as his middle name my wife's maiden name Weston. In his late adolescence Matt berated us for this unfair largesse to the younger child: "How come," he wanted to know, "Daniel has a middle name and I don't?" Now that he has started to publish articles and books, he has chosen further to truncate the name he considers already foreshortened: he styles himself simply "Matt Cohen." His brother, who has the luxury of three names, finds them too cumbersome; his professional name as a musician is simply "Dan Cohen." These are two separate approaches to name-modification. Matt's decision, I think, is a populist gesture; he is plain Matt Cohen. Dan's simplification is one approach to the entertainer's desire for name recognition: the Arlington Brugh/ Robert Taylor approach rather than the Artist Formerly Known as Prince approach. A name can be a political statement. It can also be a professional one, a shingle hung out to let the world know we are in business.

In our modern assurance, we feel we construct our own destinies and make our own worth. It is therefore disturbing that names are always *already made for us*. We cannot control them so easily as we would like. Even the child who enters the arena of childhood society with an innocuous name may suffer from having an odious nickname forced upon her: she may start out Cynthia Wentworth and become Tubby. Nicknaming outside the family, of course, is the same issue made large: it is loss of control, not just to the elders of the family but to the herd. Nicknaming by one's peers is a leveling process. Naming, as the God of Genesis demonstrates by giving the power of it to Adam, is control, whether Adamic, parental, imposed by peers, or our own efforts at renaming or reclaiming names.

23

But naming is also somehow magic, in that names ought to be a key to the secret nature of their bearers, but though we may believe in this connection, we don't know how to tap into it when we commit the powerful act of naming a child.

We cling to names: even though there may be dozens or hundreds of other Michael Cohens out there, I still prefer my name to a number, however unique. We are also annoyed when someone gets our names wrong. James Thurber begins his little gem of an essay, "The Case Against Women," with this sentence: "A bright-eyed woman, whose sparkle was rather more of eagerness than of intelligence, approached me at a party one afternoon and said, "Why do you hate women, Mr. Thurberg?" Thurber is subtly anticipating one of the points in his case — the fact that, as he says in the essay's hyperbolic mode, women "almost never get anything exactly right." For instance, they say, "I have been faithful to thee, Cynara, after my fashion" instead of "in my fashion." But of course, in addition to this point, which is really that women aren't necessarily attentive to the things *men* think are important, he is also showing how annoying it is when someone doesn't get one's name right.

In a time when privacy grows increasingly scarce, people want to put their names on pens and shirts and beer glasses and boats and cars. There is a whole industry devoted to "personalization", which means writing one's name on things. We want to keep our names guarded, but we also want to spread them around. We go along with the proliferation of access names and screen names, which, unless we have names so unique as to be bizarre, must be changed to hybrids of letters and numbers, initials and parts of our names. We are not at all sure that our names are the right ones for us, but we do want other people to pronounce and spell them right. Why do names arouse such conflicting impulses? I don't know. But I do know that we all, the Tohono O'odham, Cynthia Wentworth, the Artist Formerly Known as Prince, and I, have in common our stubborn insistence, our tedious refrain. They call me Marty but my real name my real name my real name is Michael Cohen.

3. A Place to Read

I should like the window to open onto the Lake of Geneva — and there I'd sit and read all day like the picture of somebody reading.

– John Keats, letter to his sister Fanny, March 13, 1819

I've always suspected that the proliferation of Starbucks and other specialty coffee stores was caused by people who don't actually like coffee. Your true coffee lover will drink anything that passes for the stuff, from the so-weak brew that the traditional coffee shop or breakfast parlor serves to the stark and acrid distillate of the pot that's been on all day at the office. My drawing teacher, who carried a coffee mug around through his long day of teaching studio classes, told me once that if you liked coffee enough you could learn to like it cold, by which he meant tepid, after it had lost its pot heat — and so he did. The Starbucks customer craves something different from the pure cuppa joe — something involving technology and materialism, a touch of Europe: coffee bisque or coffee soda, cream of coffee or tincture of coffee. The real coffee lover doesn't need frills or even fresh. The sailor long at sea will stoke his pot with the coffee he's got, weevils or not. The Arizona prospector carefully keeps his dried-out grind from blowing away in the wind as he boils the water for his java.

I want to make a similar point about the true reader and places to read. It's the dilettante reader who needs the light just so over his favorite comfy chair, who buys the "serious tools for serious readers" that one catalogue so pretentiously describes: tilted editors' stands and rolling adjustable reading carts like hospital meal tray holders. The really serious reader doesn't even bother sitting down for half her reading: you'll find her hunched over the counter in the kitchen reading a newspaper story whose headline ("Librarian Robs Coffee Shop") has caught her eye, standing between the aisles in the bookstore going through the entire first chapter ("The Little Black Dress") of Nancy Smith's book *The Classic Ten*, or, her book hand through the strap on the subway, clutching her purse with the other hand, she stands swaying through her whole commute, even though seats have opened up around her, while she finishes Turgenev's *Fathers and Sons*.

There is no chair, bed, or couch space in my house where I haven't spent time reading. I read on the living room couch, sitting or lying down, at the dining table, in a Backsaver chair in the bedroom, at my study desk, in bed, sitting on the toilet. I also read while riding as a passenger in cars, trains, and airplanes, in restaurants when I am eating alone, in the waiting rooms of doctors' offices and Jiffy Lubes, as well as in the local public library and the college library where I used to teach.

The miracle about reading and space, of course, is that reading transcends space: it takes you away. You don't think about where you are because, your consciousness being the most important part of you, you aren't *there* in the most important sense. Malcolm X makes the point strikingly in his *Autobiography*, when he describes learning to read in prison: "Every free moment I had," he says, "if I was not reading in the library, I was reading on my bunk." And the experience transforms his sense of where he is: "Months passed without my even thinking about being imprisoned. In fact, up to then, I never had been so truly free in my life."

Barbara Holland says we don't have to be actually reading to be transported by a book; it is enough to *have read* it. She describes a child—I take it to be her, though she makes the gender male—who has had this experience in school. The child is bored and begins to lose touch with the room and the teacher's voice. "He might struggle to hang on to reality," she writes, "but in a matter of minutes his essence was sucked from him and remolded into Huck on the river, Kim on the Grand Trunk Road, Jim Hawkins on the *Hispaniola*, leaving but an empty husk at his desk." I am quoting from her essay "Books", in the delightful *Endangered Pleasures: In Defense of Naps, Bacon, Martinis, Profanity, and Other Indulgences* (1995), where she argues that books can "move permanently into one's head and construct their own space" so that we can go inside our own heads to open a door to go elsewhere.

Reading Rooms

When I think about the power of reading to transform place and the way real readers read anywhere, I can't help but have mixed feelings about the idea of "reading rooms", places designed specifically for reading. On the one hand, I know their importance in research libraries: the reading room keeps the books contained while they are in use as well as makes the researchers as comfortable as is practical. One can't have scholars taking First Folios out onto the Folger's front steps so that they can read in the spring sunlight. I have spent many hours in the reading rooms of research libraries, the huge circular, domed reading room of the old British Museum library and the humbler manuscript reading room that was just down the hall, the elegant Huntington Library reading rooms, one with a view of the magnificent gardens, the Folger's reading rooms with their ersatz Elizabethan comfort, and the New York Public Library's reading room, with its two rows of long tables, each table with a bookstand at its outside end, nearest the shelves of reference works.

On the other hand, the most familiar library reading rooms for me are not places where special collections are in use but rather places meant for students to read and study, and about these rooms I have my doubts, the result of many distractions over many years. Two such reading rooms, where I've spent more time than in any others, are in the library at my alma mater, where I studied for nine years and three degrees, and the one in the library at the college where I taught for twenty-seven years. Both were built in the twenties, as parts of their respective campuses' main libraries, and both have been relegated to other purposes now that newer main libraries on each campus have superseded the originals.

The University of Arizona's library reading room when I went to school there — now a reference reading room for the Arizona State Museum — is a huge chamber at the front of the building, facing the palm-lined avenue that leads through the campus. On that front wall are nine tall arched windows, each with a horizontal break two-thirds of the way to the top. The windows go up to within two feet of the twenty-foot ceiling. Smaller windows, eleven in all, surround the sides and top of the arches. There are low bookcases under the windows, and six-foot high bookcases between them. On the inner wall, a six-foot high bookcase runs the whole length of the room, which is 125 feet long and 40 wide. Panels run up another five feet above the bookcases, and the rest of the wall, up to the twenty-foot ceiling, is plastered. All the bookcases and paneling are of cherry, but the window moldings are made of Mexican *amapa*, a dense, pretty hardwood from the Sonoran highlands south of the border.

I was reading Virgil for my sophomore Humanities class. We only had to read the first four books of the *Aeneid*, but I didn't seem to be getting on too rapidly; I had been at it over an hour and I was only about four hundred lines in. Aeneas and Achates had just encountered Juno's temple in the new city of Carthage. With them, I was entering the temple

> where steps of bronze led up
> To a threshold bronze-inlaid and doorpost bound
> With rivets of bronze and great bronze hinges creaking
> In the bronze doors

That's a lot of bronze, I thought. I looked around the room. The only bronze — or perhaps it was brass — was in the chandeliers: they were suspended by chains and had a brass or bronze band from which the lamps projected. Into the brass band was engraved the Greek border decoration called a... what was it called? That'd be on the Greek architecture quiz.

Above the chandeliers, two large encased beams ran the length of the room, intersected by nine cross beams and paralleled by smaller beams. The background between beams was painted in verdigris (Juno's temple was quickly receding), and the beams themselves were beige, decorated in pale salmon and grey paint. The bays defined by the crossbeams were outlined in dark green, and each of the smallest beams had a design that looked like a low relief Corinthian capital repeated three times, each repetition divided from the next by a salmon panel. The crossbeams had a salmon reversing design like a shield, all set into the beam a few inches, and the inset surrounded by a carved molding. The larger long beams had at each end an X made of leaf forms, and in between a shell and several flourishes in salmon and gray. The whole ceiling joined the wall with two sets of moldings, the first of gold squares separated by double salmon lines; the lower molding, a carved plaster one of gilt acanthus leaves.

I studied that ceiling more than any passage in my books. It struck me at the time, and it still does, that the room's size and its ornateness was really no help to the purpose for which it was intended. Its settled, classical order soothed the eye that strayed from the book. Nothing in the room rebuked the inattentive reader.

Twenty years later, I was in the reading room of the Forrest C. Pogue Library on the campus of the college where I taught, Murray State University. It was evening, and the arched windows appeared to be glazed in black. Here, too the rounded-arch windows stretched up eighteen feet or more, though the room was a more comfortable size and shape: fifty feet by eighty feet, with a ceiling twenty-four feet high. Here square pilasters separated the windows, which looked out on three sides of the room. Ionic capitals and carved plaster bosses topped the columns, with the university's shield below plasterwork drapery. This ceiling was coffered, with the three large central coffers having octangular bosses from which hung single chains that each separated into four lamp supports, with single, hemispherical globes surrounded by a brass ring decorated with the Greek decorative border called... a *meander*! That's it!

I realized suddenly that Eavan Boland was in the middle of a poem whose first half I had missed while I stared at plasterwork and brass. She was reading "The Latin Lesson" about studying Virgil as a child in a convent school:

> My forefinger crawls on the lines.
> A storm light comes in from the bay.
> > How beautiful the words
> > look, how
> vagrant and strange on the page

I like Eavan Boland and her poems; I didn't like to be giving her half my attention, but it seemed somehow to be the ironic fate of reading in this reading room, a room where we in the English Department did not go to read, but to be read to by visiting poets and novelists. We held readings here because it was a beautiful space. But beauty has its own presence and life.

In the poem, the girl studying Virgil finds in the pages a presence and a savage life that she dare not take back to the unsuspecting nuns who demand civility and gentility of their charges: she wonders how

> before the bell
> will I hail the black keel and flatter the dark
> boatman and cross the river and still
> keep a civil tongue
> in my head?

I saw the little girl, meeting Virgil earlier than I had, taken out of her reading space into his particular underworld. I took my eyes from the pretty, neoclassical plasterwork and listened to the poems.

You Are There

Anne Fadiman has an essay on what she calls "You-Are-There" reading — perusing Yeats in Sligo or John Muir in the Sierras, being on the site the author describes as you read the description. She writes that "reading books in the places they describe" thrills readers like her "because the mind's eye isn't literal enough for us. We want to walk into the pages." I am not a big devotee of You-Are-There reading, but I have felt the way a historic place resonates with significance when I have read about it beforehand. Civil War battlefields generally leave me cold, for instance, but when I visited Gettysburg after reading Michael Shaara's *The Killer Angels*, the sight of Little Round Top made the hair on the back of my neck stand up.

My one memorable experience with You-Are-There reading involved Robert Louis Stevenson's *The Silverado Squatters* (1884). This little book is an account of the 1880 summer he spent with his new bride at Silverado, an abandoned silver mine on the slopes of Mount St. Helena above the Napa Valley. Stevenson is an ideal observer at an ideal time. He has all the journalist's and novelist's skills of observation as well as the opportunity for satiric description that only comes when the point-of-view is from outside the observed society. But he was also mellowed out with the bliss of a honeymoon. Some of *The Silverado Squatters* is so idyllic one hardly thinks of his transatlantic trip in steerage or the long overland journey (he just hints at these hardships), the camping in a roofless abandoned mine shed open to the rain and the cold, northern-California nights (he insists it never rained), or the almost daily climb up the slopes of Mount St. Helena with lungs already compromised by tuberculosis (he does not mention the climb or the consumption). But Stevenson's eye is undimmed by sentiment when he skewers small-town pretentiousness by cataloging streets in the infant settlement of Calistoga, for instance.

> The railroad and the highway come up the valley about parallel to one another. The street of Calistoga joins them, perpendicular to both--a wide street, with bright, clean, lowhouses, here and there a verandah over the sidewalk, here and there a horsepost, here and there lounging townsfolk. Other streets are marked out, and most likely named; for these towns in the New World begin with a firm resolve to grow larger, Washington and Broadway, and then First and Second, and so forth, being boldly plotted out as soon as the community indulges in a plan. But, in the meanwhile, all the life and most of the houses of Calistoga are concentrated upon that street between the railway station and the road. I never heard it called by any name, but I will hazard a guess that it is either Washington or Broadway.

I read *The Silverado Squatters* during a visit to the Napa Valley and the surrounding wine country in 1999. I was on sabbatical, it was spring, and, since I was neither planning to teach Stevenson nor to write about him, I could savor his prose in complete relaxation, helped of course by the complementary savor of some Napa Valley wine. During the lunch we ate in downtown Calistoga, I don't remember noticing whether our restaurant was on Washington or Broadway. I do remember opening the book during lunch and reading the passage above to my companions. We also laughed at another point in the book where Stevenson captures a typical American practice of appropriation; he transcribes a sign he sees: "The Petrified Forest. Proprietor: C. Evans." Later that afternoon, still in Calistoga, we went to see a much advertised geyser, a small-scale Old Faithful that goes off every twenty minutes or so. It was enclosed in a high fence and charged for as if it were a circus act. Unlike Stevenson, I neglected to note the proprietor's name.

The oldest vineyard Stevenson visited in the Napa Valley was owned by a Mr. Schram, who called his wines Schramberger. The modern owner makes only sparkling wine; it is called Schramsberg. Over a glass of it I opened *The Silverado Squatters* and read Stevenson's prediction that the wines of California will ultimately prove more valuable than the silver mines. He is not impressed with the wines now, he writes, but they will have their day:

> Those lodes and pockets of earth, more precious than the precious ores, that yield inimitable fragrance and soft fire; those virtuous Bonanzas, where the soil has sublimated under sun and stars to something finer, and the wine is bottled poetry: these still lie undiscovered; chaparral conceals, thicket embowers them; the miner chips the rock and wanders farther, and the grizzly muses undisturbed. But there they bide their hour, awaiting their Columbus; and

nature nurses and prepares them. The smack of Californian earth shall linger on the palate of your grandson.

Reading these lines, I had the taste of a current California wine on my tongue, I was standing on "Californian earth" and Stevenson seemed to be talking directly to me across the generations. Anne Fadiman may have had such a feeling in mind when she wrote her essay recommending reading while on the site the author is describing.

Like the Picture of Somebody Reading

A perennial subject in art is the young woman absorbed in a book. There are paintings, drawings, and engravings of the subject by Picasso, Gwen John, Edward Hopper, Sargent, Matisse, Morisot, Cassatt, Vermeer, Fragonard, Rembrandt, Albert Moore, Fredric Leighton, Monet, Manet, Winslow Homer, Renoir, and many others. In a painting in London's National Gallery, Rogier Van der Weyden shows the Magdalen reading, sitting on the floor of a fifteenth-century room, her back propped against a carved cabinet, oblivious of the ordinary life going on around her. Manet and Monet and Cassatt and Corot show women reading in beautiful outdoor surroundings which they ignore. Fragonard's *Young Girl Reading*, in Washington's National Gallery, shows a girl beautifully dressed and coiffed, propped against a large pillow, engrossed in a tiny book she holds with her right hand with its little finger delicately curved in the air. Thomas Rowlandson's 1784 watercolor shows a woman similarly absorbed, but wearing an outdoor costume and sitting on the ground, propped between the legs of a young man.

One of the most striking representations of this subject is a 1952 photo taken by Jock Carroll of Marilyn Monroe, studying lines from a script book. She lies on a double bed with both pillows propped under her head. Only sheets and one rough wool blanket clothe the bed. She has the sheet tucked under her arms; what we can see of her shoulders, upper arms, and neck is bare, except for a thin chain around her neck which seems to have a pendant just out of sight under the sheets.

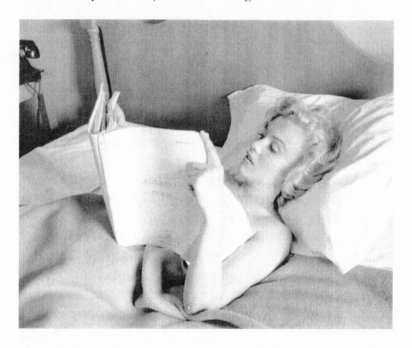

The curve of the headboard is partly visible, as is a telephone on a stand. The place appears to be a hotel room.

She is reading the script of *Niagara*, a 1953 movie in which she starred with Joseph Cotten. She has not progressed very far into the pages. Except for the telephone, the headboard, the blanket and some shadows, everything in the photograph is very light: the white script cover, white sheets and pillowcases, and her fair skin and blonde hair.

Her lips are slightly parted as she reads. Her face in repose has a little-girl openness and an appearance of continual surprise. The oversize script binder in her hands might be a book too large for her—a little girl with a big book. But that impression is only momentary. Her face and arms and the book do, however, take over the scene; the rest of the room recedes even as it does in her consciousness as she reads. The photo might be merely a modern version of the subject of the woman reading: like Fragonard's young lady reading she lies propped on pillows; like the readers of John Sloan, Giovanni Boldini, and Robert Delaunay, she has no clothes on (though discreetly covered by sheet and blanket). In all such scenes a woman is taken out of her surroundings, is oblivious of them, no longer in a physical place but projected into a place of the imagination, a story space. As reader she enters the text, and its adventures are happening to *her*. But here the movement from the place of reading into the reading is even more pronounced: Marilyn Monroe projects herself into a text written to invite her. She is thinking about how she *will be* the character in the story. As she reads the hotel room recedes, with its harsh light, rough blanket, and intrusive photographer. She will not only enter the story in imagination but will enter physically into another level of the story, into a dramatic realization of it, and she is thinking both about how the story appeals but also about how she will play it.

Marilyn Monroe was known as an eager reader, and her make-up man told the photographer that often she would conceal a book of philosophy or psychology within the pages of her script, so that when she seemed to be studying lines, she was really reading Freud or Schopenhauer. Eve Arnold's famous 1955 photo shows Monroe reading James Joyce's *Ulysses*. In Jock Carroll's photo, the reading material really is a script, the script of her first starring role, the one that took her from the place of a minor sexpot starlet right into the role in *Gentleman Prefer Blondes* that made her a superstar. But there are quite a few images of her reading, whether scripts of movies that were escape fantasies for millions, or "serious" books that she hoped would take her out of the world of the bimbo into a place where *she* was taken seriously. Norma Jeane wanted out of the little Los Angeles house, Marilyn Monroe wanted out of the bimbo box into the serious actress's dressing room, Miss Monroe the superstar wanted out of her depression, or just out. She was always trying to get out of one place into another, almost always in a rented room, looking for the script for the next day.

The Carroll photo has all this specific reference to Marilyn Monroe while retaining the generic interest of the subject. Representations of women reading show up through six centuries of European art. In these images, the male gaze can rest on the subject without fear of being seen to stare, while the woman's attractiveness is the more attractive for not being consciously employed. The subject retains some of the paradoxical features of its original, Magdalen-reading-a-sacred-text model: the woman is a sexual object but taken out of reach by her activity; you can look but don't touch. She is in a very special reading room to which there is only visual access.

Reading on the Move

When I was a child I used to read while riding in the back seat of our car on trips across the country. I was so small that sometimes I read curled up on the floor; on one cross-country trip I read most of both testaments of the Bible in the back seat, in various positions that make my old back ache just to think about now. Since then I have done a lot of reading on the move.

I will always associate my reading of *Zorba the Greek* with the voyage I read it on, steaming across the Atlantic in 1963's stormy October on the little North-German Lloyd freighter *Breitenstein*, which had Kazantzakis's book, in English, in its tiny library. (The tiny selection was actually heartening: here was a library where one could actually read all the books. Usually, the figurative place occupied by the serious reader is the center of Jorge Luis Borges' infinite library; in every direction stretch shelves of more and more books one ought to read.) It was my first trip to Europe, persisted in despite of my parents' wishes, and Zorba's free spirit spoke to a part of me that wanted to be even more of a rebel. At one point I remember taking the book onto the open area just forward of the cabin superstructure, during the most violent of the three days of storm we had during the seven-day passage. The ship, less than six thousand tons, labored in huge waves, pitching forward, rolling (at each meal the food dishes on the tables slid, even though the cloths were doused to keep things in place) and yawing as the stern settled into the troughs of waves. The crests of the waves seemed at least as high to me as the ends of the cranes latched against the freight derrick between me and the bow. It was a wild ride and a wild read; I remember shouting into the screaming wind, imitating Zorba, the adolescent's ideal hero.

When I told my friend Dave Mail about reading *Zorba* on the ship, he volunteered that on his first trip across the Atlantic, he had read Edgar Rice Burroughs' *Tarzan of the Apes* on the taffrail deck of the *Seven Seas*, tearing out pages as he completed them, letting them flutter away in the wind. "It was easier to keep my place that way," he said.

Automobiles, for most of us who have the luck not to be always driving them, provide the most common moving reading places. I've read hundreds of maga-

zines and newspapers, thousands of student composi-
tions and tests, and not a few whole books in the passenger seats of cars.

Recorded books on tapes or CDs allow drivers to
"read" as well. Of course, it's not quite the real thing.
Real reading absorbs the whole attention. I remember
listening to tapes of Jane Austen's *Persuasion* as I drove
from Kentucky down to New Orleans once, and at first
I had to keep rewinding the tape whenever I realized I
had lost the subtle phrasing of a sentence by a moment's
shift of my complete concentration to the road. Austen,
the most conscious of stylists, is probably not the best
choice of authors to listen to while driving, and it took
me some time before I could make the compromise to
give her only so much of my attention as I could spare
from threading the network of small county roads that
took me from my home to the nearest Interstate.

Is it possible that one day we will be able to read
books while driving?

We seem to have an increasing tolerance for distraction: cell phones, map displays, complicated CD
and AM/FM tuner controls and read-outs—all compete for our attention while we are driving. Someday I
would not be surprised to see in cars a moving text display, something like the TelePrompters political can-
didates use to project their speeches onto transparent
screens through which they can see and be seen by the
audience. Drivers of the future could probably learn to
change focus rapidly back and forth from the projected
lines of text to the road beyond.

A Place to Read

"Reading transcends space," I wrote earlier; "it
takes you away." Well, yes and no. An odd relation
always links reading, readers, books, and space. The

strangeness affects the space around readers and the space around books. In the individual case, the reader's absorption in the book conjures the reader's envelope *into disappearing*. In the case of libraries, the reading act conjures a space *into being*. The structure of a library — its bricks and mortar and steel shelves — comes into being as a function of the contents of its books. A library's objects are arranged by an abstract system of words, letters, and numbers such as the Dewey Decimal System or the Library of Congress Classification. The abstraction becomes an arrangement of shelves, layered into stacks. The library's physical arrangement never exactly reproduces this abstract map, of course, but a library is an architectural manifestation of an intellectual order, a place that comes into being because of the act of reading.

But around the reader, the act of reading can also do the opposite of annihilating space: it can bring into being a multiple consciousness of place. Late in his memoir *Speak, Memory*, Vladimir Nabokov reminisces about reading *War and Peace*: "I had read [it] for the first time when I was eleven (in Berlin, on a Turkish sofa, in our somberly rococo Privatstrasse flat giving on a dark, damp back garden with larches and gnomes that have remained in that book, like an old postcard, forever)." What a picture of somebody reading! Nabokov has just mentioned Prince Andrew Bolkonski's jejune enthusiasm for war and "Tolstoy's dismissal of the art of war" — two poles of feeling in the book, calling up the sweep of the book's action and its movement from Bald Hills, the country estate of the Bolkonskis, to the battlefield of Borodino and the metropolis of Moscow. Tolstoy's fictional spaces are superimposed upon a real landscape he knew and historical events finished before he was born. Reading this picture and map, Nabokov coyly dominoes every reader's inner and outer worlds into the room (somberly rococo) with its sofa (Turkish) and the garden (dark, damp, with larches and gnomes) into which he peers — and is it the larches and gnomes "that remained in that book, like an old postcard" or the whole scene inside and out? The scene makes clear how we do not lose the immediate world in reading; we acquire others and fix (like a photo or a postcard) the one we're in. Like a picture of somebody reading.

4. Meat and 3 Veg

Tell me what you eat and I will tell you what you are.
— Brillat-Savarin, *The Physiology of Taste* (1826)

If you want a plate lunch in the little college town of Murray, Kentucky, you go to Rudy's on the court square, Roberson's Hihburger on Route 121 south of town, or Martha's on Route 641 north of town. Be warned that each is a smoking establishment and the strongest drink you can order is the iced tea — you will be asked whether you like it "sweet or unsweet". The plastic glasses in which the tea is served are clear for the unsweetened variety, tea-colored for the sweetened, so that the waitresses, who often refill your glass, will know which to give you.

Rudy's has eighteen square Formica tables and six stools crowded at the counter. In wintertime it's a tight squeeze for the diners, who are often large anyway, and their fat winter coats. Behind a separate, higher counter is the cash register, and behind the register is a cold case with the day's pies. The popular meringues and cream pies are probably Rudy's main claim to uniqueness among Murray's plate-lunch parlors. These days there is always a sugar-free dessert choice as well. When Rudy's is full at peak breakfast and lunch hours, eighty people may be masticating or waiting. The wait is short: usually a plate lunch arrives within about four minutes after the waitress takes the order. If the owner, Dana Anderson (Rudy sold the place to her years ago) is in the tiny kitchen cooking, the wait may be even shorter. And no one is sitting around after eating; if you're finished and you don't move, you will be shoo'd out. Charging two dollars for breakfast and five for lunch, Rudy's depends on turning over those eighteen tables and six stools.

If you choose a plate lunch — and most do — you can put together a vegetarian meal of four selections, but the usual combination is one meat and three vegetables. The meats at Rudy's include honey-glazed ham, liver (beef liver rather than calves' liver, served with onions and gravy), and meat loaf, but the standard fare is chicken with a sage dressing or with dumplings, tenderloin — that is, pork tenderloin cooked in a flour batter, roast beef (dry-roasted and pulled or cut into small pieces), and "country-fried steak", which turns out to be ground beef cooked "chicken-fried", that is, with a flour and egg batter. Some variant of the "country-fried" or "chicken fried" or "breaded beef pattie" shows up on all the menus in town. On Fridays there are fish options, again cooked in a flour or corn-meal batter. The salmon patty is always there on Fridays, and two-piece or three-piece fish dinners with hush puppies, fried potatoes, and slaw — a fish and three veg without the choices. I like to go on Fridays, not for the fish, but because liver and onions is always on the menu that day, and one of the vegetable choices is lima beans. My wife abhors this combination, so I don't get it at home.

Rudy's has one vegetable selection you won't see anywhere else in town: the tomato relish. Tomatoes are chopped with onion and jalapeño peppers and served in vinegar with a little sugar to take the bite out. Otherwise, the vegetables are much the same as those served elsewhere in town. The whole-kernel corn, creamed corn, and lima beans are much as they come from the can. With other varieties of peas and beans — purple hull peas, black-eyed peas, pintos, and white beans — a little salt pork or ham usually seasons the dish. Mashed potatoes, always called "cream potatoes" or "creamed potatoes", show up on the menus of all these restaurants every day, and are served with gravy. Buttered carrots, green beans, beets served hot, cabbage wedges—all of these are cooked up with some sugar, although the waitress puts a bottle of vinegar down on the table with the cabbage. Vinegar is also served with turnip greens, though these are not usually sweetened. Sauerkraut is not a usual menu item in this part of the country, but occasionally will show up with Polish sausage among the meat selections. Sweetness is the dominant flavor, followed closely by salt. And in season you will often see fried green tomatoes on the menus of these places.

The odd thing about the plate-lunch restaurants is their sameness. There are small variations in standard dishes, and there are specialty dishes to be found at each restaurant, but otherwise there is a kind of uniformity that McDonald's quality-control guardians would find enviable. Of course there are differences. Roberson's Hihburger (no one now knows why the "g" was left out of the name) and Martha's feature barbecue every day. The barbecue in this part of the country almost always is pork, though a few diners also offer barbecued mutton. Unlike the Texas wet barbecue, slathered with viscous red sauce while cooking, the meat in Tennessee and Kentucky is dry-roasted in small smokehouses behind the restaurant or in cookers made of steel drums. It's pulled apart when done and served with pepper sauces of varying piquancy. The typical barbecue sandwich comes with clumps of the tender meat on a hamburger bun, often topped with slaw. At Roberson's you can also sometimes get a barbecued pork chop as your meat selection on the plate lunch. Martha's serves barbecued ribs as well.

Roberson's offers the cole slaw familiar in all these places—chopped cabbage with a little carrot in a creamed, sweet sauce—along with two other choices: vinegar slaw, which is chopped cabbage with a little chopped bell pepper in diluted vinegar sweetened with sugar, and vinegar slaw with Splenda, for the weight-conscious or the diabetic. Music plays in Roberson's—a radio station playing country music. Roberson's register is behind the counter in this crowded room with eight booths around the exterior, large-windowed walls and only four places at the counter.

Martha's, the newest of the plate-lunch places, was started in 1986. Martha is a well-kept blonde woman in her fifties with short hair and a fondness for gold jewelry. The local AM radio station, WNBS, broadcasts a morning show from this restaurant, and two tables on the highway side are marked off for this purpose. Here a roll of paper towels on an upright holder in the middle of each table acknowledges that the ribs and the fried chicken will need to be eaten with the fingers. You can get lasagna for a meat choice here, or ham shank; a selection not usually seen elsewhere is smoked turkey, which often appears on the menu. The "smothered pork chop" is covered with a milk gravy with mushrooms. Martha's is the most spacious of the plate lunch places, with 15 booths and another dozen tables—in all, space for over a hundred people. The kitchen is also the largest, and your plate lunch comes back from it almost as soon as your order has been placed.

Murray residents eat at all three restaurants, but they have favorites, and the customer base at each place depends on its location. The clientele at Rudy's always includes some lawyers and clerks from the nearby courthouse and municipal or county offices. You will also see a sprinkling of downtown merchants, a stockbroker, and the eastside farmers who began going there in the fifties, when Rudy's 99-cent breakfast was the draw.

Roberson's patrons are primarily workmen from the plumbing shop, the lumberyard, the car-repair garages, and the other businesses of the "industrial" southwest corner of town — but southside farmers also come in, as do a few young families. The college people — staff and faculty alike — are more likely to show up at Rudy's or Martha's than Roberson's, and few students frequent any of these restaurants, probably because the nearest one is a mile from campus and the crowds in all three have little in common with students in age or interests. Kids who grew up in Murray still come of course, and sometimes bring their friends to see a little bit of local color. Martha's group is the most eclectic because it's on the main north-south highway through town and draws an ever-new population from the transients. She also has regulars among the northside merchants and the county people who live north of town.

Wherever you go, your meat and three veg will be served on a plate divided into three sections, made either of plastic (Martha's) or crockery (Rudy's and Roberson's); the third vegetable will be in a small bowl placed with the meat in the largest section of the plate. You will be offered a choice of rolls or cornbread, or one of each. In the Murray plate-lunch parlors, dessert is a vegetable — that is, you may choose one of the day's featured desserts as one of your three vegetable choices. It is important, I think, during the short wait for your plate to be set in front of you, as you set your taste toward the meal to come, that you remember that dessert is a vegetable. At Rudy's it will be Jell-O or a cobbler or, occasionally, strawberry shortcake that's listed with the vegetable choices on the plate lunch menu. The featured pies cost extra.

Finally, of course, the plate-lunch places *are* the town and embody its features: a love of predictability, a surface sweetness free from irony, a resistance to change, a dash of originality in a barrel of the familiar, an attitude about food that it should be cheap, plentiful, and accompanied by tobacco but not alcohol.

5. My Hypochondria

And the prisoner released from the cave, will he not fancy that
the shadows which he formerly saw are truer than the objects
which are now shown to him?

    – Plato, *The Republic* (4th century B.C.)

I.

The black-and-white image on the little TV screen was
at first formless and incomprehensible, but, explained
by the ultrasound technician, it resolved itself—with
some imagination on my part—into a stylized heart.
First a spasm in the top third of the screen, a blossom
of light spreading and dissipating, then a spasm at the
bottom, larger, stronger. I saw it as an abstract pattern
and then suddenly I saw it as a heart. And then, as
suddenly, I saw it as *my* heart, at the same moment I
felt the tiny thump in my chest seem to grow louder
and more palpable; I could feel and see my heart beat.
My heart. It beat slowly, suppressed by the digitalis
my doctor had prescribed after my episode of dizziness, an effect of disturbance in my heart's rhythm.
Now the rhythm was very regular, though as if in slow
motion: lub DUB... lub, DUB. This "lub, DUB" is the
conventional transcription of the heart's sound heard
through a stethoscope. To me, the sound was more personal: "heart BEAT... heart BEAT" or "my HEART... my
HEART." I was hypnotized, unable to look away. This
trance did not seem to surprise the technician, who had
probably seen it before.

The inside of the body is another country. Invaded
by major surgery or injury and forced to show its

secret organs, the body's interior shocks us and makes us wince. We don't ignore the fact we have bodies or think of ourselves as wholly mind; rather, we think of our bodies as those parts we can keep an eye on—the exterior. The outer body is easier to take: we worry about its bulges, sags and wrinkles, but at least we can keep an eye on them, and we have the illusion of controlling the body's appearance through exercise, diet, shaving or refraining from it, hair dyes and dos, make-up, jewelry stuck in the skin or tattoos inked into it. We avoid the knowledge that we might *be* those slippery, bright, impolitely-colored, soft and moving shapes, those stark white bones.

The ultrasound machine shows the inside of the body without invading it with flak or scalpel. The machine shows a black-and-white picture on a TV monitor, a picture formed something like a radar image, by reflected radiation just beyond the frequency of waves we can hear as sound. The blood-red field of the operating theater or the trauma room is absent in ultrasound. Its stylized and abstract image speaks to the observer, saying "I am my organs", but saying it in the metallic accents of a machine. Ultrasound images can be cause for joy and wonder when the device helps expectant parents view their unborn, growing children. Much used in obstetrics, ultrasound is the safest way to get a good look at what's going on inside the body. Heart specialists like ultrasound also. With it they can watch the heart at work.

I watched my own at work for perhaps twenty or thirty beats, thinking about this amazing little pump that had always worked so regularly, so constantly and unfailingly without my thinking about it—a little pump upon which the whole of my consciousness and my life depended. Then the conviction began to grow in me that this constancy and this perseverance was more than amazing; it was magic. And I don't believe in magic. I saw the little pump as just another piece of connective tissue, like the ones that had been failing me over the last year. "It cannot continue to do that," I thought, "minute after minute, hour after hour, a hundred thousand times a day. It will stop. Maybe in five more beats." I watched and counted five more beats. It didn't stop. "Maybe in a thousand, but it will stop. And when it stops I will die." The fear in this thought was so palpable I expected it to be somehow visible in the image on the screen: the beating would suddenly speed up or pause. But its chemically-slowed regularity persisted. lub... DUB, lub... DUB. "Not while they're watching," it seemed to say. "Tonight, in your bed. On the road in the country. Somewhere far from help."

The melodrama of these thoughts galled me, chagrined me, but knowing the thoughts were overdramatic did not stop them. I had always been either offhand at the prospect of death or unconscious of it. I knew I would die someday, of course. I had come close to death a few times in cars and once or twice in planes, close enough to be aware that my surviving in those instances was a chance that could easily have gone the other way. I knew intellectually that I would not live forever and that I could die at any moment. But the thought was not often present, and when it was it did not oppress me or obsess me.

Staring at the monitor now, though, watching my beating heart, I found the thought of my imminent death both oppressive and obsessive. It was also unwarranted. My heartbeat was now normal after an episode of atrial fibrillation caused by exertion and overindulgence in alcohol on a hot day, dehydration, and a consequent imbalance in electrolytic fluids—solutions of sodium and potassium in the blood that facilitate the electric impulses of nerves to heart muscles. This imbalance, once corrected, was not likely to return. After a week or so of precautionary medication, the doctor would take me off the digitalis. "You're fine," he would say, "and your heart is sound." But, starting on that day in the ultrasound room and for two years afterward I was convinced, if my heart fluttered or I had a momentary chest pain, that I was going to die.

What was even worse, I began to extend this fear beyond worries about my heart. If I wheezed a little, I was sure that a lung had collapsed or I had lung cancer. If my urine came slowly I knew it was prostate cancer. None of my organs were safe. I had become a hypochondriac. Whenever I felt a twinge, a momentary shortness of breath, or an accelerated heartbeat, I was convinced it was the onset, not just of sickness, but of mortal disease.

II.

The modern definition of *hypochondria* is an abnormal anxiety about health: the sufferers fear or actually believe that they have a life-threatening disease: a cough means lung cancer, a stitch in the side means a heart attack, a headache means a brain tumor. The doctor's reassurances to the contrary are either not believed or else are effective only until the next ache, pain, or minor symptom convinces the sufferers anew of the onset of the mortal threat they fear. The word *hypochondria* has not always been so specific in its denotation; at the time Shakespeare was writing, the word was another name for what the English variously called the *spleen* or *melancholy*. This was a much more general ailment, corresponding to a series of neurotic disorders we would now call by the blanket term *depression*.

In 1621, Robert Burton wrote a long treatise, *The Anatomy of Melancholy*, about this general malady. In

Burton's scheme, hypochondria is merely one of the three bodily areas of origin for melancholy, which he says can originate in the mind, in the body as a whole, or just in the organs of the abdomen—the *hypochondries*, which literally means *below the cartilage of the ribs*. For Burton, melancholy is a disease, and he gives us the theory of disease most popular during his time, which said that it results from a lack of balance among the body's constituent fluids or *humors*: blood, phlegm, yellow bile or *choler*, and black bile or *melancholy*.

Burton describes the symptoms in great detail, and I was not far into his description before I began to recognize the very complaints my friends and my family members had told me about when their doctors had diagnosed them with "clinical depression". Burton's melancholics are sleepless, anxious—sometimes with rapidly pounding hearts and hot or cold sweats. Their main symptoms are fear and sorrow. The fear is without cause, and may be a conviction that something disastrous is about to happen or that they will suddenly die or will suffer some disease or injury. Their fear may be only of disgrace or embarrassment, but in any case it is a disabling fear—paralyzing. Sorrow or sadness is also a constant symptom. Also without cause, it makes the sufferers restless and guilty and troubled, irresolute, indecisive, and robbed of the will to act. Suspicion, the readiness to take offense and to misconstrue what others say or do, is also a symptom, as is what Burton calls *humorousness*; we would call it quick mood changes. Burton's vocabulary does not include phrases like *mood swings* or terms such as *paranoia*, but he effectively anatomizes the whole spectrum of ills we call depression, anxiety disorders, including hypochondria, and panic attacks.

The causes of melancholy, both supernatural and natural, get a lot of space in Burton's *Anatomy*. Supernatural causes include not only evil spirits, but God himself, imposing the disease for his own inscrutable reasons. As cures, Burton recommends prayer, the care of an honest physician, moderation in diet, good air, exercise, "mirth and merry company." His concluding advice is *"Be not solitary, be not idle."*

The identification of hypochondria with depression continued into the eighteenth century. James Boswell, the biographer of Dr. Samuel Johnson, wrote about his own low moods and periods of helpless inactivity — in other words, his depression — in a series of columns for the *London Magazine* where he signed himself as *The Hypochondriack*. Boswell wrote one of these essays every month for almost six years, from November 1777 to August 1783. He thought of the writing as therapy, combining the discipline of the deadline with the knowledge that self-examination might bring.

Boswell understands the sufferers' belief that they are very ill as only one of many manifestations of the complaint called hypochondria, which he equates with melancholy, "the spleen, or vapours." For Boswell, the main quality of hypochondria is irresolution, sometimes extreme languor, together with agitation about what one should be accomplishing while idle.

Before the eighteenth century was over, doctors had begun to separate from general "melancholy" those neurotic symptoms that relate to the sufferers' beliefs about their health, and a new name for the complaint, *hypochondriasis*, began to be used among some physicians. This term did not replace the words *hypochondria* and *hypochondriac* among lay persons, but all these words came to denote the condition and the person who suffers from ungrounded fears about health, rather than the whole complex of symptoms described by Burton and Boswell as part of the melancholic's complaint.

Of course, even before he was called by that name,

the hypochondriac was among us. Jane Austen depicted such a character in Mr. Woodhouse, the title character's father in *Emma* (1815). Mr. Woodhouse worries
and guards not only his own health, but is solicitous
about that of everyone about him. He warns the neighborhood children, in vain, not to eat the cake at a wedding celebration, and, even when he is host of an entertainment, would much prefer that his guests eat a bowl
of warm gruel rather than the more enticing foods he
and his daughter have laid out for them. Mr. Woodhouse has a mild form of hypochondria, redeemed
from the neurotic extreme by his optimism: he knows
that danger and disease lurk everywhere, but he believes that they can be defeated by constant vigilance.
Austen did not call Mr. Woodhouse a hypochondriac;
the term she used was *valetudinarian* — one concerned
for his health. But the latter word could be confusing,
since it was sometimes used of a person whose concern
was real rather than imaginary — one in weak health. In
any case, *hypochondriac* has superseded *valetudinarian*,
a word hardly ever seen these days.

Hypochondria has recently begun to be treated as a
variety of depression, with some of the same tools used
for the more general illness. According to Dr. Ingvald
Wilhelmsen, a Norwegian physician and psychiatrist
who specializes in hypochondria ("Norwegians are
melancholy," he says, "it's very dark here most of the
year."), it is "excessive health anxiety", as common in
men as in women, and found in people of all ages. Dr.
Wilhelmsen recommends a talking cure, with behavior-
al modification and building up of the patients' awareness of their tendency to move from a minor symptom
to suspecting the worst possibility for its origin. He
has also tried his patients on a drug, one of the class
of medications known as selective serotonin reuptake
inhibitors, that are marketed in the United States under

such names as Paxil and Lexapro. Wilhelmsen was featured in a *Wall Street Journal* article in 1996. The article's fairly jocular tone probably reflects a general attitude among the public about hypochondria. It's a difficult malady to take seriously, unless of course one has it.

The good news, had I consulted Dr. Wilhelmsen when I had my problem, would have been this: the easiest case of hypochondria to treat is one that comes on later in life rather than being a part of one's psyche from childhood. These later onset cases often result from major life changes or critical events; the death of a close friend is one triggering factor. Wilhelmsen's characterization, which I read some years after the events I am describing, seemed to fit my experience.

III.

As I turned forty-four, my back began to give me trouble. At first it was a pain in my left hip, then pain down the back of my leg and numbness in my foot, then spasms in the muscles of my lower back. What had been annoying became incapacitating. I could get relief in the beginning by walking or by lying down; only sitting or standing still were painful. Then the pain would only relent when I was flat on my back. Finally, it was constant.

I had the ruptured disk in my spine repaired later that year in an operation called a laminectomy. The surgeon cleaned up the disk and enlarged the hole in the backbone segment through which the sciatic nerve exits; the ruptured disk's pressure on the nerve at this point had caused the pain. The procedure sounds simple, but since it involves the spinal column and therefore all the nerves that serve the lower part of the body, it has risks. Late in the evening before my surgery, my back surgeon showed up in my room flanked by two nurses, to explain the risks and get my signature on a release form.

The nurses were there as witnesses. My doctor, a small neat man ("Do *you* have back trouble?" I had asked him. "Oh no," he said, "I'm a little guy.") went through the list of possible bad outcomes, a litany that included loss of feeling in one or both feet or legs, paralysis, loss of control of bladder or bowels, and loss of erectile function. The litany ended not, as might have been appropriate, with *ora pro nobis*, but with death. I signed the form, they instructed me to "get a good night's sleep", and they turned out the light as they left. To my surprise, I did sleep. The surgery went well and my recovery proceeded without any setbacks. At the end of the year I was playing golf with no thought about the pain of a few months before.

Then in the spring of the next year I woke up wheezing one morning. A little walk winded me, so I went to my doctor. He was about to send me home with some asthma medicine when I mentioned that my chest also hurt a little. He listened again with his stethoscope before sending me for a chest x-ray that revealed my right lung had collapsed.

If air gets into the chest cavity outside the lung, either through a wound or, as in this case, through a small hole in the lung itself, the air pressure outside the lung prevents it from inflating. No disease or injury had happened to make my lung deflate; my doctor called it a "spontaneous" collapse, a word that evoked for me all the wrong associations of Whitman's "spontaneous me": insouciance and unrehearsed charm. Charming it wasn't. A collapsed lung is treated by first cutting a hole in the side of the chest between two ribs. Through this hole the surgeon pushes a tube, which is then hooked to a suction pump. The pump evacuates the air inside the chest cavity around the lung, allowing the lung to inflate. The tube remains connected to the pump until the hole in the lung seals itself; in my case this took three days. Then the surgeon pulls out the tube, puts a purse-string suture around the wound, tightens it up, and covers the whole thing with a dressing, all the while trying to keep more air from leaking back through the wound into the space around the lung.

I recovered quickly from the collapse and the repair (the chest-tube procedure leaves a scar like a stab wound — is in fact a stab wound), immediately resuming my usual activities with no ill effects and no further lung problems. To this day no one knows why my lung collapsed.

In the fall of that year my oldest friend called to say he had colon cancer that had spread to his liver. He talked lightly about an experimental drug program he was entering, but he had no illusions — and would not suffer his friends to entertain any — about the fact that he was going to die. And until his death a year later he maintained an attitude that, while not always cheerful, was curious, open, and calmly accepting of the stages of his disease and his death. "Get a colonoscopy," he said, and I did, but the clean bill of health the doctor gave me was not a surprise to me; at that time I did not have the hypochondriac's conviction that the news would always be bad. The onset of my hypochondria came suddenly with my heart problem the summer following my back operation.

June that year was even hotter than usual, and I liked to have a cold beer or two — occasionally three — as I played golf. One day I felt particularly exhausted after a round, and I lay down to take a rest. My heart started to flutter. It felt as if it was trying to get out of my chest. For seconds at a time it beat very fast, and I grew dizzy. No pain accompanied the fluttering and the dizziness, but these symptoms were alarming enough for me to call my doctor, who said, "go to the emergency room." There the attending doctor told me I had atrial fibrillation, a misfiring of the electrical impulse that begins the heartbeat in the atria, the two chambers at the top of the heart. In fibrillation the heartbeat begins but is never completed, so little or no blood is pumped. The faulty heartbeat speeds up to try to compensate, and may reach 150 beats per minute, more than twice the normal rate. After a test revealed low potassium levels in my blood, the doctor ordered an intravenous solution to increase the potassium, which is one of the substances — sodium is another — that makes blood, an electrolyte, able to transmit electric current and thus nerve impulses. I remained in the hospital overnight, hooked to a heart monitor with an alarm in case the heart rhythm worsened. During the night, as I slept, the rhythm "converted" or changed back to its normal, efficient pumping action that starts in the top of the heart and is completed by the stronger muscles of the ventricles at the bottom. The doctor let me go home the next day, but I had to return later in the week for a stress test, which is an electrocardiogram or graph of the heartbeat taken while one is exercising on a treadmill. The doctor also ordered an ultrasound test of my heart, and it was while I watched the ultrasound waves imaging my beating heart, as I have written above, that my hypochondria started.

I had gone through a year-long demonstration of what can go wrong with heart, lungs, spinal chord, and bowels. A healthy response would have been gratitude for my escape, when my friend with colon cancer was not so lucky. Instead I became fixated on the body underneath the surface. I felt I *was* these organs and bones. I began to study medical encyclopedias and books, which had the predictable effect of increasing my fears. Hypochondriacs seek relief through knowledge about disease, but instead of relief they find symptoms to worry about ("Is this red spot the beginning of a rash?") or alarming facts ("Every 44 seconds, someone dies of heart disease.")

For two years I stayed tensely alert to the inevitable failing of my body. A dozen times a day I checked my breathing to see if it was shallow or wheezing; were my lungs okay? If my stomach griped I could envision the cancer growing there. Every time I went to a doctor I would be reassured for a few days or weeks; then with the next heart flutter or tiny stitch in the side the conviction would return that something was fatally wrong. And then, even more suddenly than it had come, this neurotic delusion went away.

IV.

I flew in to Salem by way of Seattle to do a lecture and workshop at Willamette University with Bill Braden's students. Bill had liked my book on *Hamlet* and had invited me for the Willamette University Senior Seminars in the Humanities. Willamette is a small liberal arts college and these Senior Humanities Seminars focus on a single text. The instructors invite a scholar/ teacher who has done work on the text to come for a few days to meet with the seminar, confer individually with the students on their paper topics, and give a public lecture on some topic related to the seminar.

On the last day of my stay, Braden took me to lunch just before he drove me to the airport. Midway through our sandwiches and microbrewery beer, he told me a shocking story. The year before, during a routine physical, the doctor had taken him into his office and announced that Braden had a fatal stomach cancer that would kill him within months. Braden lived with this judgment for several weeks before telling his parents, who then told him he had a congenital thickening of the stomach wall that showed up on x-rays as disease. "When the doctors realized their mistake and corrected their diagnosis," Braden said, "they didn't even apologize."

My reaction to Braden's story, a mixture of horror that it was possible, compassion for his ordeal, and effusive congratulations on his escape, was followed by more self-centered musings: *"What does this mean for me?"* I thought. What does the possibility of a mistake in an unanticipated sentence of death (for Braden had no hint or fear of what the doctors would tell him) mean to a hypochondriac, who is always anticipating such a sentence? My happiness for his reversal of fortune Braden may have thought oddly overdone from an acquaintance of a few days. And the extent of my joy was hard for me to explain, even to myself.

On the way back, flying from Salem to Seattle on the first leg of my return home, I was sitting next to the window as we flew through a thunderstorm. I thought about Braden's story. At the moment he told me that the stomach cancer diagnosis had been revealed as false, I felt in my own stomach a realization that will sound like a commonplace to anyone who has not felt depression or hypochondria: *there is good as well as bad fortune.* The hypochondriac thinks of contingency and chance as all going one way. One's health can only get worse; news about it will always be bad. But Braden's story showed that even the diagnosis one fears isn't always right.

There was little lightning but a lot of rain, and as I stared idly out the window at the rain streaming straight back along the Plexiglas, I became aware of a flashing white light coming from below me. Apparently a powerful strobe light was mounted on the bottom of the fuselage where I could not see it, but as I pressed against the window looking down toward the source of the light, I saw below me fat raindrops hanging in space, motionless, each time the strobe flashed. I blinked and drew back from the window. Rain rushed across the outside surface, and just inches away it whipped past the window at several hundred miles per hour. I put my face to the window and looked down again. Globules of water were suspended in the gray night. Unmoving. None of the drops was a perfect sphere, and as I looked closer they seemed to quiver slightly in the image that persisted after each flash. For minutes I stared at these unnervingly arrested raindrops below my window, until the shower stopped and we began to descend into Seattle.

I continued idly staring out into the night until the flashing strobe began to light the ground rushing up at us from the obscurity below. I thought about the unnaturally still raindrops, and the strobe's strangely revealing light. I thought back to that other strange illumination of the ultrasound, revealing my heart in hiding. These machines show us entrancing pictures, but they do not tell the truth. Rain falls; it doesn't hang suspended in space. The heart beats, as it should, in secret. There are reasons why our skin is not transparent like some jellyfish's. The skin is the rightful limit of our concern: we can attend to it, clean it, clear it of nits and mosquitoes, but we gain no advantage from being able to see our secret hearts. We cannot see time arrested any more than we can see the future and be certain that it holds ill fortune. I thought these thoughts and knew that, for now at least, my hypochondria was gone.

6. Men in Uniform

As they observe, "What does it signify how we dress here at
Cranford, where everybody knows us?" And if they go from
home, their reason is equally cogent: "What does it signify how
we dress here, where nobody knows us?"

   – Elizabeth Gaskell, *Cranford* (1851)

One of the many reasons I am glad to be male — right
up there with never having to deal with menstruation
and usually being able to get my carry-on out of the
overhead compartment by myself — is the clothes. I
don't feel the urge to express myself creatively through
clothes, which is just as well, since men's clothes are
now and have been for the past couple of centuries
devoid of originality. Like most men, I am content to
wear the same sort of costume worn by other men of
my age, habits, and condition. We very often put on
what amounts to a uniform.

Wearing the same clothes as other men and not
thinking about making a statement about myself with
clothes means essentially never having to think much
about them at all. Are they clean? Do I have them on
right-side-out? That's pretty much the extent of the
questions I have to ask myself about how I'm dressed
most of the time. Dave Barry talks about "Basic Guy
Fashion Rules" and says one example is "Both of your
socks should always be the same color, or they should
at least both be fairly dark."

I was interested, though, to run across a little book
by Nancy MacDonell Smith called *The Classic Ten: The
True Story of the Little Black Dress and Nine Other Fashion Favorites* (2003), which discusses parts of women's

wardrobes that have some staying power against the changing force of fashion. She talks about pearls, about cashmere sweaters and trenchcoats and other perennials of the female wardrobe. I was so struck with this book that I stood in the aisle at Barnes & Noble while I finished the first chapter, on the Little Black Dress. This fashion favorite interested me because it seems to deny the advantages women's clothes have over men's: namely, color, and variety. The LBD, as Smith abbreviates it, works by what she calls "refusal". It gestures away from the privation and sadness of mourning—if not the ultimate origin of the Little Black Dress, mourning costume is at least an allusion made by it. The absence of color is a kind of deferral—and a promise—of gaiety.

If several women are wearing little black dresses, they invite a comparison, which will quickly go beyond the dresses themselves to the women wearing them. The LBD partakes of the male tendency toward uniformity while it reveals a paradox of uniform dress. On one hand, sameness of costume is a place to hide; I won't stand out because I'm dressed peculiarly. On the other hand, if all of us are dressed alike, you may look at each of us more closely to see the person under the costume, and thus frustrate any wish I may have for anonymity.

All of these thoughts led me to ask whether a list like Smith's could be made for men's clothes. Are there "classics" of male dress? Certainly there are a few dress combinations that keep showing up.

Consider what is called by most people the "tuxedo", but is known in tonier circles than I frequent as "evening clothes". From one point of view, how dull: every man at the ball or the banquet is dressed the same! But from a more reasonable point of view, how brilliant! I need only pack one suit for all the formal gatherings on the cruise. Notice in this respect that the tuxedo trumps the Little Black Dress; while Audrey Hepburn got away with wearing the same one in scene after scene of *Breakfast at Tiffany's*, it is unlikely any woman I know would show up two nights in a row wearing the same frock. But I need never worry whether my costume is going to look expensive or exclusive enough to keep me in countenance when I'm given the once-over. (Note that the "I" here is only a convention; I don't actually own a tuxedo.) I can attend all the inauguration balls all week without bankrupting myself on my getups. And the tuxedo works for every degree of formality from the Tinyville Charity Ball to the White House Gala.

These days, tuxedos can be bought ready to wear, jacket and pants separately sized to your proportions and needing no alterations. Where once the shirt, with perhaps a detachable front and certainly a separate collar, had to be boiled in starch, these days the customs have softened. All that a man needs is a shirt that accepts studs and links, the black, pleated sash with the delightful name *cummerbund* (from a Hindi word meaning *waistband* — apparently a modification of British military dress uniform first adopted in India), and mastery of the slightly tricky art of the bow tie.

Some misguided souls sell or rent tuxedos in garish reds and blues, as well as cummerbunds and bow ties that are striped or plaid. These clowns' costumes are fine for high school proms, but they are made by people who have missed the whole point of what formal dress does for men: it makes them look alike, which means they can be comfortable while their penguin-like uniformity serves to highlight the spectacular gowns of the women they accompany. The classic here is a dark coat with satin or grosgrain lapels, black pants with a stripe of similar material along the outside seam, a

white shirt with links in the French cuffs and studs replacing the usual front and collar buttons, a black bow
tie, a black, pleated cummerbund, and black, usually
patent leather shoes.

The tuxedo was introduced as formal wear in Tuxedo Park, New York, in 1886. The tobacco-rich Pierre
Lorillard, imitating the Prince of Wales (who liked to
wear a short dinner jacket rather than the traditional
tailcoat) brought the style to the States, and his son
Griswold and a few friends wore the short, dark coats
to a formal autumn ball where all the other men were
in white tie and tailcoats. To increase the effect, they
wore red waistcoats. The waistcoat that was worn with
tails and originally with the tuxedo eventually gets
simplified to a pleated sash—just an abbreviation of
the vest—that we know as the cummerbund. The moral of this history may be that men's clothing evolves
toward the simpler and more comfortable, although,
as we shall see, it may also move toward other evolutionary purposes.

At the other end of the scale of formality, another men's classic consists of khaki chinos, a blue button-down shirt, and a pair of brown loafers without
socks. The premier college outfit of the 1960s, this costume was the showcase of the 1980 satire *The Preppy
Handbook*. It survives because of its simplicity, comfort,
and versatility: I can wear this outfit (this "I" is not
a convention but really me) into most restaurants as
well as into the grungiest student bar. Nowadays this
combination may identify me as being of "a certain
age" because it was most favored by the generations
before and after the Baby Boomers came of age, but I
see younger men wearing it, too. The advent of wash-
and-wear fabrics made the chino and button-down
combination even more convenient, especially since
a few wrinkles seemed to be acceptable, although the

combination can be sported with knife-edge creases and perfectly ironed shirt front. As with most male uniforms, there is room for minor differences in appearance and huge differences in price. The younger and more affluent will favor plain front chinos and Brooks Brothers shirts, often without pockets. Older wearers may have pleated pants and shirts from Lands End or a mass-market department store. The shoes may vary as well: although Bass Weejuns were the usual choice in the earliest decades of this costume's history, coastal wearers modified it with the cut-sole boat shoes developed by Paul Sperry in the 30s and known as TopSiders.

Other variations can include a polo shirt or a Hawaiian print shirt instead of the button-down, and the chinos can turn into shorts in warm weather. The biggest variation came when technology addressed the growing popularity of running and suddenly brought us shoes more comfortable than any we had known before. But some would say these variations, especially the running shoes, which involve putting on *socks*, so dilute the original combination as to take it out of uniform status.

If you add a blue blazer to the chinos and loafers, you attain the first stage of male formal dress. Add a tie and you can get into even the snootiest restaurants. Of course, you may have to put on some socks.

Notice how incursions keep being made on the next level of formality: tuxedos invade the white-tie-and-tails compound; blazers and slacks intrude on the sacred territory of the suit. Suits, though, have not been relegated, like tails, to near extinction. Many consider the suit the quintessential male uniform.

A man's suit, according to Anne Hollander's book *Sex and Suits: The Evolution of Modern Dress* (1994), reveals the articulation of the body beneath but at the same time pads and regularizes it in order to flatter the wearer. Its form is thus a compromise between fashion — that is, a clothing dynamic that is considered to gather favorable attention to its wearer — and comfort that adapts to body shape. But the suit has not always made this quiet compromise. Seventeenth-century examples sported huge ballooning upper arms, codpieces, and large knickers that attenuated into tights, the whole colored in the brightest of hues and made of expensive fabrics. Costumes in Shakespeare's day flattered and flaunted the wearer's sex while they allowed also for display of money. The suit has evolved toward a more modest — and some of my gay friends insist a much more boring — presentation of the outlines of the male form. As it has simplified its lines, the male suit has also moved toward concealing rather than revealing, by cut, color or attachments, what rank in society its wearer occupies. Anne Hollander, again, says that this ideal begins to be realized in the eighteenth century with Beau Brummell, who "was known to have wished his clothes unnoticeable." Thus, she continues, "the Neo-classic costume was a leveller in its time."

Modern suits *can* reveal to the careful observer when they cost fifty times more than the one J. C. Penney sells; the markers of sartorial not-so-conspicuous consumption make up one of the obsessive themes of Brett Easton Ellis's *American Psycho*. But a careful eye is required to appreciate the lawyer's five-thousand-dollar Armani. Male suits now admit a fairly limited number of styles, cuts and tailoring — so limited that the features which respond to fashion are relatively inconspicuous: the width of lapels and of ties, the flare of a pant leg. The biggest thing that has happened in male suit fashion over the last fifty years is the strange movement of the double-breasted suit into and out of fashion. But double or single-breasted as the case may be, the suit is not primarily designed to assert itself as fashion, as attention-gathering. And thus the male suit-wearer has a psychological advantage over the female, however closely hers may model on the male uniform. If I take my one suit to the three-day convention, I have no hesitation about wearing it several times, merely changing my shirt and tie. For women, the suit is a costume rather than a uniform, a day's choice rather than a three-day wardrobe.

Nowadays my warm-weather uniform consists of a tee-shirt or cotton-knit golf shirt with shorts, wash pants, or, less often, Levis. The key to this costume is that no part of it needs ironing. Oh, the wash pants would probably look better if they were ironed, and once in a while I send them to the cleaners to be pressed. Mostly, though, I wash them and let them dry on a hanger or stretched on the frames that Vermont Country Stores sells for this purpose — the lazy man's way to get a crease in his trousers. In the summer I often wear boat shoes with no socks, but all year my normal footwear is white cotton tube socks and walking shoes. The whole getup is the extreme version of comfortable and convenient.

What I *do not want* is an emotional investment in the clothes I wear, and many men feel the same. A man's version of Ilene Beckerman's *Love, Loss, and What I Wore* (1995) is unthinkable except perhaps for gay men: straight men do not ordinarily associate their affective lives with their clothes or shoes. I tried the experiment of remembering what I was wearing the day I got the news I'd passed my Ph.D. exam, the evening when I proposed, or the night when my first son was born. Nope, nothing there. Of course I must have been wearing *something*. If I were very inappropriately dressed for an occasion, I might recall the offending outfit, but as we have seen, one of the governing principles of male sartorial design is making inappropriate choices hard to blunder into. A tuxedo will do for every formal evening, whatever the occasion; a suit takes over for the next half dozen levels of formality, but often a blazer and slacks will do for all of these as well. And so it goes all down the clothesline, making the wrong choice as rare as possible, right down to the socks or no-socks decision. There are deliberate, huge overlaps in "suitability". In my clothes nightmare, I show up naked at the ball, not wearing the wrong shoes.

7. The Victims and the Furies

During the point in the jury selection process called *voir dire*, prosecutors and defense attorneys question prospective jurors and, everyone hopes, have their questions answered truthfully. This provides an opportunity for both sides to size up those who will decide the case. The following are actual questions I heard while waiting to learn whether I would be empanelled on a jury in a recent capital murder case in the 6th U. S. District Court (I was not selected for the jury).

"Have you or any of those close to you ever been the victim of a violent crime? "

Yes, my father was shot to death when I was a young child.

"Do you have any religious or moral beliefs or convictions that would prevent you from imposing a death sentence?"

No.

"Do you feel that society has a right to impose the death penalty for especially heinous crimes?"

Yes.

My answers may puzzle some readers. If I'm not morally opposed to the death penalty—if in fact I would be willing under certain circumstances to impose it—why do I take a position against it?

I oppose the death penalty not because it is morally wrong, but because it is ineffective and dangerous. Furthermore, it doesn't deter criminal behavior, it's more expensive than life imprisonment, it's unsure, and it's sold politically and implemented widely in ways that pander to racial bigotry. Worst of all, it threatens

71

through another sort of pandering—this time in the name of "victims' rights"—to undermine the very basis of justice.

It is this last point that I wish to discuss in detail.

Twentieth-century journalist and social critic H. L. Mencken pointed out that the shocking or degrading nature of the death penalty is as irrelevant to most people as is the fact that it doesn't deter others from heinous crimes. The most important aim of capital punishment in Mencken's opinion is revenge. The immediate victims of the crime and society as a whole, he says, want "the satisfaction of seeing the criminal actually before them suffer as he made them suffer. What they want is the peace of mind that goes with the feeling that accounts are squared." As usual, Mencken is useful in getting us to admit what really motivates us, regardless of what we claim the justification for our acts might be. Revenge—the ancient spirit embodied for the Greeks in the form of the Furies—is the real justification for killing people judicially, and we will not think clearly about the whole process until we admit it to ourselves.

Vengeance seems to be society's strongest reason for embracing the death penalty, regardless of what name we give it. And it does not bother most people that it might actually be in conflict with other justifications, such as deterrence. As Gary Wills points out in a June 2001 article in *The New York Review of Books*, juries that impose the death penalty for horrendous crimes "reflect more the anger of society," while they "fail to make the calculations that we are told future murderers will make."

Perhaps the awakening of the Furies, this "anger of society" without coolness or reflection, is justified by heinous criminal acts. Perhaps the egregiously cruel, the wantonly violent (Timothy McVeigh once referred to the nineteen young children who died in the Murragh day-care center as "collateral damage"), and the sadistically homicidal acts of murderers may justify society's thirst for retribution and reprisal. Never mind that the New Testament revises that so-called biblical injunction of taking an eye for an eye. "When the injury is serious," writes Mencken, "Christianity is adjourned, and even saints reach for their sidearms."

Well, if society is making the judgment on the basis of the wrong done to it as a whole, then it would be futile to cry out against the judgment. But if, as seems more likely, what is increasingly happening is a failure to distinguish between the interests of society and the interests of the immediate victims, then the whole purpose of a judicial system is being subverted. We need to remind ourselves what exactly justice is and what it replaces, lest we revert to a pattern where vengeance drives all reaction to crime.

Unfortunately, we can't review the origins of law and courts directly because they are lost in prehistory. We can, however, pay attention to what literature and myth tell us about courts of law replacing the rule of revenge.

The Oresteia is a group of three plays written by the first great Greek tragic dramatist, Aeschylus, in 458 BCE, when Athens was at the height of its civic and artistic glory. The plays move from a world where violence is answered by violence toward a newer, more civilized world in which Athenian society votes as a jury on guilt and penalties for crimes; law replaces vengeance.

In these plays, Aeschylus gives us a myth about the origin of the justice system, the very system the West later inherited. In the process he shows how society has to absorb and neutralize the force of individual vengeance and the desire for revenge if there is ever to be justice. It's a lesson that has been forgotten in the current furor over victims' rights. We cannot isolate the victims of crimes as somehow being more entitled to retribution; justice means that society as a whole enacts punishment for a crime against society.

In Aeschylus' narrative, which runs through the three plays, the returning Greek king Agamemnon is murdered by his wife Clytemnestra out of revenge for his sacrifice of their daughter, Iphigenia. Clytemnestra is later killed by their son, Orestes. But when Orestes is pursued by the Furies, who wish to avenge his murder of Clytemnestra, Apollo and Athena enter the action to invent the Athenian court system. They try Orestes and eventually acquit him.

The Athenian justice system offers a way out of the cycle of revenge. For Agamemnon and his house—as for any society ruled by blood vengeance—the past has captured the future. Aeschylus expresses this brutal constraint in *Agamemnon*, the first play of the trilogy, with imagery that confounds past, present, and future.

An example comes in the opening, as the old men who make up the chorus recall an event that took place before the Greeks sailed to Troy to avenge the kidnapping of Helen. The priest Calchas, summoned by Agamemnon to figure out why the lack of winds is keeping the avenging Greek fleet from sailing to Troy, reads a sign: two eagles devour the unborn babies of a rabbit. Calchas says the child-protecting goddess Artemis, angered at the coming slaughter of Troy's children, retards the inevitable progress of the Greeks toward their Trojan conquest until the Greek leader serves up his own child as a sacrifice. In the verses of the chorus, the eagles' feast of unborn young is mixed up with the sacrifice of Iphigenia, the slaughter of the Trojan Queen Hecuba's children and all the young innocents at Troy, and the outrages that began the curse on Agamemnon's house — namely, the murder and cannibalism of children perpetrated by Agamemnon's father, Atreus, and his great-grandfather, Tantalus.

Tantalus served his son Pelops to the gods for a meal and was punished by being unable to grasp the grapes he continually reaches for in Hades. Atreus killed the children of his brother Thyestes and then served them to Thyestes for a meal, calling down Thyestes' curse on the family. Agamemnon sacrificed his own child to appease Artemis and eventually was murdered by Clytemnestra in revenge. Orestes, the son of Clytemnestra and Agamemnon, then killed Clytemnestra in revenge for his father. In the imagery of the plays, the present and the future are both controlled by the past. Time is a prison where the deeds of the past are continually recommitted with different victims.

The myth of the cursed house of Atreus illustrates two important truths that must be grasped before any society can achieve even the beginnings of justice. The first truth is that murder is always a family affair in the broadest sense: all of society is implicated and diminished by the killing of one of its members. The Hebrew culture conveys this truth by having the first murder that of brother by brother. But the Greek myth contains another truth not seen in the Genesis tale of Cain and Abel. Though in both stories the first murder is divinely punished and the punishment is continued torment, in the Greek story the human reaction to murder is to answer it with another murder. This mortal vengeance seems a natural response, but as the myth makes clear, it invites further vengeance. The second truth, therefore, is that murder answered by revenge inspires revenge in its turn. To this cycle of retributive vengeance there is no end. In the words of Mohandas Gandhi, "An eye for an eye makes the whole world blind."

Aeschylus' subject is the problem of vengeance and the origin of the justice system to answer it. When Clytemnestra kills Agamemnon, she invokes "the three gods to whom I sacrificed this man", which she names as "the child's Rights... Ruin... and Fury." She also says she is moved by "our savage ancient spirit of revenge." In fact, Aeschylus darkens her motivation by having her glory in her adulterous affair with her lover, Aegisthus, and in the power the two of them wield over the city-state of Argos.

Because in Aeschylus' story Clytemnestra's son and killer, Orestes, will eventually escape the cycle of vengeance, the tragedian depicts him not as driven by the Furies, the primitive spirits who are the essence of revenge, but as the instrument of Apollo. The god not only threatens Orestes with consequences if he fails to act—ostracism, physical pain, and psychological torture—but he promises his own protection. As Orestes kills his mother offstage, the chorus says that the son is guided by the goddess of justice, Athena, and that "Apollo wills it so!"

Though Orestes is pursued by the Furies after he kills Clytemnestra, Athena and Apollo arrange it so he is put on trial and acquitted. This requires Apollo to claim that Orestes' act was Zeus' justice, and the jury comes in at half for conviction and half for acquittal. By breaking the tie and casting her vote for acquittal, Athena symbolically inaugurates the practice that tie votes will result in acquittal. Finally Athena, who knows the Furies aren't just going to go away, promises them a place in the depths of earth and renames them as kindly spirits, the Eumenides.

Though this allegory isn't, itself, an argument against capital punishment or against the belief that a death should be justified with a death, the message is that the wrong done in crime is felt by the whole com-

munity. But in this it is felt more impersonally and coolly, objectively and rationally, than how the victims and
their immediate families feel it. And when the punishment is removed from the hands of those immediately
wronged, it is made a less personal matter. The crime
is punished rather than the victim avenged. And the
punishment isn't designed to make the victims feel better or experience "closure". It is society that is wronged
and thus society moves to punish the crime. Until it
does so — coolly, not in the heat of vengeance, deliberately rather than swiftly — there will be no justice, only
an endless cycle of crime following crime.

A routine practice in sentencing hearings in U. S.
courts these days is the hearing of "impact states" —
testimony from the relatives of a murder victim. Their
voices cry out because the victims' voices cannot. The
burden of those voices is almost always the same: the
parents, siblings, and spouses of victims won't achieve
peace, or in today's jargon, "find closure", until the
murderer is dead and as absent from his or her family as is the victim. Victims' rights organizations have
sprung up all over the country, loudly complaining that
the courts protect the rights of the accused and even the
convicted at the cost of those of victims and the families
of victims. It is hardly possible to exaggerate how important and widespread the victims' rights movement
has become. According to Scott Turow, whose experience as a member of Illinois Governor Jim Ryan's task
force to reform the state's capital punishment system
changed him from an advocate to an opponent of the
death penalty, "the national victims' rights movement
is so powerful that victims have become virtual propri-
etors of the capital system."

It hasn't always been thus. Within the last several
decades, the U. S. Supreme Court reversed its earlier

decisions that argued that victims and their families had no role in the process of deciding guilt and passing sentence. As recently as 1987 the court said that statements by victims and relatives weren't constitutionally admissible at a capital sentencing proceeding.

During the 1988 presidential debates CNN's Bernard Shaw asked Michael Dukakis, who was perceived and promoted by Republicans as soft on the death penalty, what he would do if his wife Kitty were raped and murdered. Dukakis fumbled his response, which should be clear and the same for anyone asked such a question: "I would want to kill the people responsible, of course. But what does that have to do with the question of how society should act responsibly on the question of assigning penalties for crimes? We don't give guns and police cars to people whose family members have just been killed, open the fingerprint data base to them, and tell them to find the killers and do to them what they would like to do."

Aeschylus' fable points us to the real relation that ought to exist between society and victim. We need to tell victims that revenge is *not* one of their rights. In the rule of law, society steps back at least a couple of paces from the victim's position. Society doesn't say to the victim, "go and take revenge yourself." Neither does society say, "you have been wronged and we will act as your substitute in exacting revenge." Under the law, society measures the extent of harm done to it — to the whole body of the people — and imposes punishment based on that wrong. This degree of abstraction is precisely the measure of your safety and mine from reversion to the rule of the Furies.

Abstraction, of course, makes an unsatisfactory meal for victims. My family choked on the verdict of manslaughter rather than murder that enabled my father's killer to get out of prison in a year. For victims a life is "worth" a life, which means taking one demands taking another. The Furies, too, argued at the Athenian court that their rule—older than the law, older than the gods—demanded a life for a life, and if their rule weren't honored, they further warned, the center wouldn't hold.

Society has always felt that killings in some circumstances weigh less than they do in others. When it comes to deciding punishment for the perpetrator, a life taken after its owner puts himself in harm's way—starting a bar fight, for example—doesn't carry the same judicial weight as an innocent life wantonly taken. These are simple, hard facts of society's judgments. A classic case is the crime of passion. The irate husband who kills the adulterer isn't treated in the same way as the serial killer. But if we succumb to the appeal of victims' rights, we can no longer justify the distinction.

The widow and the three children of the adulterer—I am one of those three, for this is my father's killing we are looking at here, not an abstract example—suffer his loss as acutely as any mother of one of Jeffrey Dahmer's victims suffers the death of her boy, or any husband of a wife blown up by Timothy McVeigh suffers hers. Turow, looking at the issue from a lawyer's point of view, points out that "it violates the fundamental notion that like crimes be punished alike to allow life or death to hinge on the emotional needs of the survivors."

The sudden absence of the murder victim from the lives of family and friends means that no one can say "I love you" one more time, or make up a quarrel, or, in my mother's case, ask "Why did you betray me?" The anger and perhaps guilt that goes with these cruelly truncated relations generates its own desire for revenge.

But not in my case: I am free from the anger and the guilt. When I think about what should have happened to the man who killed my father, I have a luxury other victims lack: I am emotionally insulated from the event of my father's killing because it happened when I was too young to know what was going on. Such emotional insulation is precisely what society has and needs in order to pursue justice. It isn't the business of juries to feel the victims' pain but to decide what harm has been done to society. It isn't their business to give satisfaction or closure to victims but, rather, to render swift, reliable, unbiased punishment for the harm done to society as a whole. As a victim, I would like juries and the society from which they are chosen to think not as victims do but as members of a responsible body that measures wrong by a different yardstick from emotional pain.

And, as an added benefit, taking the heat of revenge out of the sentencing process means that sentences will be fairer across lines of gender, race, and class because bias is more likely to sneak into the process the more passionately it is conducted. We need to restore to our courts the objectivity the Greeks depicted as necessary after so many generations of murder and revenge.

8. Seeing Mexico

Para bailar la bamba, se necesita una poca de gracia.

   – *La Bamba*

During spring break of my senior year, three of my college buddies and I boarded an Aeronaves de Mexico flight that took us first to Hermosillo and then on to Guaymas. Though I had grown up in Tucson, only ninety miles from the border, it was my first sally beyond the border towns of Nogales and Tijuana. In 1965, Guaymas and San Carlos Bay were still fairly primitive in their approach to tourism. Before the Pemex boom, the few tourist hotels stood next to wood shacks and even open ramadas where whole families lived. I walked the streets staring, not knowing what to make of my first third-world experience.

Nothing seemed to work in the town: the electricity blinked off for hours during the day, hot water was impossible even in the mildly tony hotel where we stayed, and everywhere the appearance of the modern reverted to the reality of open sewers, cooking done on outdoor fires of wood or any other fuel, and every child under six naked except for a dirty shirt.

And then we went to the satellite tracking station. The first manned Gemini mission had just launched the day of our arrival. Gemini 3, with Gus Grissom and John Young aboard, was circling the earth every ninety minutes. The tiny capsule, which had been only slightly enlarged from the Mercury model that carried only one man, orbited from west to east, each circle monitored by tracking stations across the globe.

In the Guaymas station, we were introduced to two young astronauts, Alan Bean and Gene Cernan, neither of whom had flown a space mission yet. They showed us how the radar picked up the signal from the spacecraft even before radio contact could be established. Conversation between the two astronauts on board the spacecraft and the ground was only possible while the orbiter was over the horizon. They called the radio connection "line of sight": to talk to the satellites you had to be able to see them. In about four minutes, the spacecraft flew from horizon to horizon. Then another tracking station in Texas would have to take over. Stations were arranged around the world so as to maintain almost continuous contact with the spacecraft.

One of my buddies, bolder than I, asked the two astronauts to accompany us to dinner, and to our surprise they accepted. We drove to a hilltop restaurant overlooking San Carlos Bay and ate cold shrimp and sliced avocados. Royal Poinciana and Mexican Bird of Paradise blossomed in the green shadiness that surrounded our covered, dirt-floored patio. We drank bottled Carta Blanca beer, not quite cold enough, and looked out at the Sea of Cortez. I wasn't at that time able to appreciate how odd a juxtaposition the space–age activities of the clean-cut astronauts made with the scene around us. But as I looked out at the bay I thought about the question that had been forming in my mind as I walked the streets: "What had I come to Mexico to see?"

***

Alain de Botton writes in *The Art of Travel* about how the novelty and uniqueness of a place to which we've traveled might well be easier to perceive if we hadn't had to bring along the same apparatus for perceiving it that we use at home. The traveler must, perforce, bring himself along on the trip, and the self decreases our ability to see the novel and exotic. De Botton goes so far as to suggest a paradox: "We may best be able to inhabit a place when we are not faced with the additional challenge of having to be there."

***

The area known as the Barrancas del Cobre—Copper Canyons—is part of the Sierra Madre Mountains of northwestern Mexico. It began to be accessible to tourists with the 1961 completion of the remarkable railroad running from Los Mochis, on the Sea of Cortez, right up to the city of Chihuahua. Northwest of the town of El Fuerte, at the point where the States of Sonora, Sinaloa, and Chihuahua come together, the railroad begins to climb in switchbacks along nearly sheer cliff faces, through a series of tunnels up to a mile long, and over bridges spanning a handful of rivers that drain into the Rio Fuerte, which flows into the Sea of Cortez north of Topolobampo. The railroad is called the "Chepe", that is, "Ch-P" from "Chihuahua-Pacifico", and before it enters the canyon that gives its name to the whole barranca region, it parallels Urique Canyon, the deepest of all the barrancas, deeper than the Grand Canyon, nearly six thousand feet from its rim down to the River Urique and the town of the same name.

My wife and I took a trip to the Barrancas del Cobre in the late summer of 2004. We flew into Los Mochis, stayed two nights in El Fuerte, and then took the train up into the canyons, from which we were driven up to the rim above Urique. Two days later we traveled farther north to another lodge on the rim of this canyon.

Urique Canyon, though deeper, is still young by comparison with the Grand Canyon. More than half of the slopes have dense pine forest. Occasionally a mountain meadow shallows out from the steeper pine slopes; here the dark green forest gives way to light green and yellow grasses. Farther from the river rise sheer cliffs with faces of a few hundred to several thousand feet of drop. On top of the cliffs, rocks crazily eroded by wind and water make shapes of tall pillars or stout mushroom caps.

The first word that springs to mind in attempting to describe the barrancas is *sublime*. If anything is sublime, one would think, these huge grooves in the earth, which must be visible from satellites orbiting the planet, are sublime. They testify to the agency of powerful forces: vulcanism, the shift of huge crustal plates riding on red-and-white hot oceans of magma, wind and rain, and flowing water working over unimaginable spans of time.

And yet *sublime* is not quite the right word. I know something of the aesthetic concept of the sublime, having written a dissertation thirty-five years ago about this idea as it developed in the eighteenth century. Yes, the hugeness and power of natural forces is a part of the sublime, but so is a special relation to human beings.

Eighteenth-century writers anticipated—some would say caused—the new appreciation of dramatic landscapes that is so prominent a feature of Romantic literature and painting, and of so much of our tourism since. These writers found natural phenomena of great size, power, and danger aesthetically appealing, a fundamental change in taste from Renaissance and medieval times. Chaucer and Shakespeare did not see any aesthetic appeal in avalanches in the Alps, ocean storms, or canyons with sheer precipices. Wordsworth did. What appealed about natural sublime landscapes was their power and how it dwarfed human pretensions. As de Botton says of the new aesthetic in *The Art of Travel*: "A landscape could arouse the sublime only when it suggested power — a power greater than that of humans, and threatening to them. Sublime places embodied a defiance to man's will."

The Barrancas del Cobre have all the ingredients of sublimity except this one relation to humans. The problem is the tin roofs. And the ladders. Salted among the high meadows and cliffs, from the ridges down to the river below, tin roofs glint like fool's gold among the rocks of the canyons. They cover the houses where the Tarahumara Indians live. Tarahumara dwellings have tin roofs now, though they had leaky wood-and-rush ceilings in the past. The change came because of Victoriano Churro, a Tarahumaran, a long-distance runner as these Indians traditionally are. Churro was taken to the United States to a hundred-mile race in Colorado, which he easily won. On his return to Mexico, he became a hero in his state of Chihuahua, and when the governor asked him what he would like as a reward, he pointed to the leaky houses of his countrymen and asked for tin roofs.

We visited Churro at his house and garden near the rim of Urique Canyon. Typical of the Tarahumara are Victoriano's short stature, prominent cheekbones, and crinkled dark skin around the eyes from the unfiltered sun of the high, clear air. So too is his quietness and courtesy. Victoriano and his fellow Tarahumara have a disconcerting habit of showing up suddenly on the roads and trails of the canyons, having come from above or below the more-traveled ways on a steeper path of their own. The Tarahumara move seasonally in the canyon — up during warm weather, down when it is cool. Their dwellings occupy every level, and some live in caves in the rock face, easily made habitable because their needs are simple. The women who make and sell baskets and other craft goods scale the cliff faces to the tourist outlooks on the rim of the canyon using rude wood and hemp ladders. Suddenly, as I gazed out over the rim, a Tarahumara woman would appear beside me, silently offering her wares, squinting into the bright sun. When I looked at the Copper Canyon panorama, I was always aware of the Tarahumara, whose presence gives scale to these picturesque scenes, like the grazing animals in the foreground of Ansel Adams' famous 1944 photograph, *Sierra Nevada from Lone Pine, California*. But the natural scene does not dwarf the human beings, any more than a large oak dwarfs its acorns; in neither case are the large and the small competing to establish the scale.

***

As I was struggling to take in the whole juxtaposition of landscape and people that was so far outside any experience I had had before, my traveling companions were making me think about the act of *seeing* itself. We were on a tour — the first packaged tour my wife and I had ever taken — and there were two couples traveling with us. They were all from Brooklyn, neighbors who had traveled together to many places in the world such as Machu Picchu and Africa and the Amazon. Jake and Rosy were in their seventies; Melvin and Bella were a little younger. They tended to make comparisons, in an oddly limited way, between what they saw in the canyons and what they knew from elsewhere. Every mesquite and palo verde — and this area of Mexico has many such small-leaved plants — got an identical response: "It looks like a mimosa."

Some tourists use their cameras as buffers — they need to take pictures to mediate between themselves and the travel experience. As Susan Sontag observed in *On Photography*, photographs "help people to take possession of space in which they are insecure." Although

Jake, one of our traveling companions, was a photographer looking for shots he could enter in the photo competition at the museum where he was a docent, Melvin and Rosy, and Jake and Bella did not push cameras between themselves and the encounter with the canyons. Instead they used talk. They had to articulate, usually in the most banal terms, what they were seeing, to assure themselves that they were seeing it. They had to talk to be there. They reported constantly on the surroundings visible to all of us, like the person riding in the car with you who feels the need to read aloud every sign on the road. Only when they were frightened were they silent, when we were jolting down the rock trails — unrailed, with improbably steep grades, not really roads at all — that were the only way of getting up to the rims of the canyons from the railroad, and along which our guides drove us in vehicles that didn't even have four-wheel drive. When the van lurched toward the edge and its passengers suddenly stilled their talk, I was reminded of a remark by Mary McCarthy in one of her essays to the effect that strong emotion like fear can make us better able to see the world, that it, like being absorbed in looking at something, makes us forget who we are for a moment, so that we are "taken out of ourselves."

Rosy and Jake came out on the trail with us for a birdwatching hike on the second day we were in Urique Canyon. They each carried small binoculars, and from the side of the trail they picked up branches sturdy enough for walking sticks. When someone pointed out a bird, each of them would lift binoculars to eyes, meanwhile letting go of the walking sticks, which would clatter noisily on the rocky trail. The bird, of course, would be spooked and fly away. This little comedy was repeated all along the trail. Before long, my wife and I dropped back away from the group and loitered behind, first a few yards, then fifty, then a hundred. When the rest of the group had gone round a shoulder of the canyon, we sat down on boulders and silently regarded the scene for a few minutes. "Listen," she whispered. I didn't hear anything. "What?" I whispered back. "No one is talking," she said. And then, as if a cue had been given, from the trail behind us, Melvin, who'd been hiking behind us, hailed us with an echoing "¡Hola!" He came up and began to describe in detail the spring and other sights he had seen along the trail, where of course we had just been.

Melvin and Bella, and Jake and Rosy made me think about what it meant to *see* what we were encountering. I wasn't making a lot of progress with my own methods, with simply allowing the experience to wash over me. Perhaps the banal, the "I am seeing a Military Macaw, and it's bigger than the Scarlet Macaw we saw in the Amazon" is an authentic approach. After all, interpretation is a mode of dissatisfaction that tries to substitute something else, a tamed intellectual construct, for the raw experience. What more could one do to turn one's attention from the outside to oneself than move toward interpretation? Somewhere between the banal and the over-interpretive and self-absorbed, there must be a state with the grace to see. *Para bailar la bamba, para ver el país, se necesita una poca de gracia.*

***

In June of 1991, two colleagues and I were in Mexico City. It was my first trip to the capital, and I was overwhelmed by the city's hugeness, its confusion, and its acrid brown air. As we drove in from the airport, my friends explained to me that millions of people lived on the outskirts of the city with no plumbing, and their excrement, dried and turned to dust, was part of the city's notorious pollution problem.

We had a full day free for sightseeing. We started in the morning at the National Anthropological Museum, an enormous complex of buildings around a courtyard with a huge fountain. We looked at colossal carved Olmec heads and turned to encounter almost identical faces in the people beside us. We looked at round Pre-Columbian vases with handles that were representations of the skinny forelimbs of some creature like a frog or a lizard, wrapped around the vases' bulging centers, each with little legs at the bottom.

We decided on an art excursion that began at the Museum of Modern Art in Chapultepec Park; large, colorful canvases by Diego Rivera, Juan Orozco, and David Alfaro Siqueiros were the highlights here. Then we moved to Rivera's studio not far away. We had lunch at the San Angel Inn across the street from the Rivera studio. While eating a variety of ceviches along with pompano and chicken with mango sauce, we pondered our next move. My friend from Berea College in eastern Kentucky, who has some strongly populist leanings, suggested we might go to Trotsky's house in Coyoacán, and then perhaps to the Frida Kahlo Museum nearby.

Trotsky's house is fairly pleasant from the outside, with a shady garden, but its vaultlike iron doors attest to his fear for his life, entirely justified in the event. The details of Trotsky's murder are not completely clear even now, but he was Frida Kahlo's lover at one time, and certainly hated by Rivera. Moreover, David Alfaro Siqueiros had unsuccessfully plotted to kill Trotsky, and Kahlo's sleeping with him may have been part of another plot in which Rivera was involved. Our day's sightseeing seemed to be developing a theme. The artists who sculpted those Olmec heads in the museum and the mural painters of the Siqueiros and

Rivera circle were not passive observers. They made the historic monuments and put their hands on the politics of their times. In the presence of their works I was beginning to feel how shallow the seeing was in my sightseeing.

We moved on to the Kahlo Museum, but it was closed for renovations, according to a sign by the door in an eight-foot wooden wall around the half block of grounds. The Frida Kahlo revival was well under way at that time, and the authorities were preparing for a sizable flow of tourists while they were also retrieving from abroad as many of Kahlo's paintings as could be bought up. A small self-portrait had recently fetched almost two million dollars at a New York auction and was now on its way back to Mexico.

A guard came to the door after we pounded on it for a few minutes, and when my populist friend offered him a thousand pesos to let us in, he said he would see and disappeared for a long time. When he reappeared he said he couldn't do it, but when the price was upped to two thousand he let us in. There were few Kahlo paintings in the rooms, and they were standing against the walls rather than hanging. But the furnishings were all still there as they must have been when Kahlo lived in the house with Diego Rivera. Their studios were filled with cans and jars of brushes, easels, lay figures and drawing materials. In the stairwell, tacked over and around the door, was a superb collection of the small antique paintings on tin called *retablos*.

We were in the bedroom, looking at the not-very-large bed and thinking about little Frida and three-hundred-pound Diego. A guard tapped me on the shoulder, motioning me into the adjoining room. On a bureau was a death mask of Kahlo done in metal, perhaps pewter. We looked at it, made appropriate noises, and turned to finish our tour of the bedroom.

The guard took my arm and led me back to the bureau. Now he was indicating a large pot or vase, almost spherical, with what at first looked like two small handles just above its equator. When I looked again I realized that it was nearly identical to one I had seen in the anthropological museum that morning; it must have been a pre-Columbian vase from Rivera's collection. Now he was tilting the vase toward himself, removing a saucer from the top, and pulling from the mouth a long dirty shawl. "¡Mira! ¡mira!" he said. Then he smiled and nodded toward the opening. We looked in. The bottom third of the vase was full of grey ash, with a few dusty lumps here and there which I realized suddenly were bone fragments. It was Frida! We were being given our money's worth—a view of this secular saint's very relics. What started out as a casual enough tour had turned unexpectedly into a pilgrimage.

***

Sometimes I think I have looked at Mexico through a picket fence made of mirrors. Glimpses of the country and the people come through, but I'm never sure how much of what I see is my own reflection. But I'll be going back, because there's always the promise that the next time I will see her as she is. Come and see Mexico—her clear coastal skies, filled with satellites flying east toward the Atlantic, her inseparable past and present, her megalopolis with dried shit blowing in the wind, her tourists staring, her canyons with their cave-dwelling aborigines, her statues, her pottery, her ashes, her bones.

9. Communing with the Dead

"So, you believe the dead don't care about the living at all?" I asked. I was in the middle of an eye-opening conversation with a Baptist from Campbellsville, Kentucky. We were riding in the back of a bus on the way to Tzintzuntzan, an Indian village in the Mexican state of Michoacán where tourists often go on the night before the Day of the Dead. In the village that night, the three cemeteries are filled with light. Family members keep an all-night vigil, covering the graves of their loved ones with candles, flowers, food, drink, and other offerings.

***

Earlier in the evening, at a restaurant in the lakeside town of Pátzcuaro, one of my dinner companions had mentioned that a student in the college group he was guiding had been hesitant to come along on this excursion to Tzintzuntzan.

"Why?" I asked, thinking how long I had been curious about what we were about to see. I grew up in Tucson, where Day of the Dead decorations—little skeletons and spun-sugar skulls—began appearing in late October, but I had never seen the vigil at the graves. Now, just hours from it, I was eager and excited. I couldn't imagine anyone would turn down a chance to see it.

"She said that the whole thing went against her religious beliefs," my friend replied.

"And her religion is... ?" I asked, imagining some strange cult.

"She's a Baptist," he said. "Why so surprised, Mike? You know Baptists don't believe in any kind of communion of the dead with the living."

"Do I?" I said. "But..." I stopped, ashamed of my ignorance. "They believe in an afterlife, surely, heaven and hell and all that. How can you believe that without believing in the dead and the living having a continued interest in each other?"

But my friends didn't give me a satisfactory answer; they just chaffed me for my ignorance and assured me that it was so. I *did* know that a big part of the Protestant schism from Catholicism was the repugnance felt by many of the schismatics at what they felt was idolatry of saints and of Mary. But the idea that the student who had complained to my friend could be a Christian and not feel that the spirits of the dead were still pulling for her—I couldn't quite square it with what I knew. Of course, what I knew was the product of a Catholic education through high school. I was a lapsed Catholic; my friends, on the other hand, had grown up in Protestant families and were converts to Catholicism. I had spent little time educating myself about the doctrinal arguments of the Reformation; I was always more interested in the art. But the notion of *some* kind of communion with the dead one had known when they were living seemed merely natural to me rather than doctrinal—Catholic or Protestant.

***

Now, jolting along the bad road from Pátzcuaro to Tzintzuntzan, I had singled out Darrell, one of the faculty members from the Baptist college in Campbellsville. As we pulled into a makeshift bus park near the village, I asked Darrell the questions that had been on my mind.

"Look," I said, "You pray for other people don't you?"

"Yes," he said.

"And other people pray for you?"

"I hope so," Darrell said.

"Well, " I said, "If someone who had prayed for you dies, is he no longer interested in you?"

"No," said Darrell. "He would either be damned or blessed, and either way he would not care about me."

***

The Indians who live in Tzintzuntzan believe, and have believed since before Cortez, that graves exhale their dead like the mists of autumn. The Catholicism that came with the Spaniards has its own version of spiritual communion with the dead, effected by prayer on the part of the living. The prayer is heard by the sainted dead, who then may intercede with God on behalf of the living. The Indians syncretized their own beliefs with those of the Catholic faith they "accepted" along with Spanish rule.

The Protestant schismatics so despised the corruption of the doctrine of communion with the dead by the Catholic clergy — the selling of indulgences, primarily — that they rejected the doctrine completely. Of course, those Protestant churches who adopt Calvin's view that the damning or saving of souls is a preordained matter can find no reason whatsoever for the dead communing with the living or vice versa. But I felt in my conversation with Darrell, as I had in my first hearing of the student who thought visiting Tzintzuntzan would be a betrayal of her faith, a strong feeling that denying any connection between dead and living was denying some universal human belief too deep for dogma. I began to be convinced, in other words, that communing with the dead wasn't a religious issue at all.

***

I do not claim to have held communication with the dead. And yet, friends and family members have come to me in dreams, many months after their deaths, to reassure and comfort me.

The circumstances in these dreams are bizarre, because I know the people whom I see and converse with are dead, and yet they are convincingly there. In the first such dream, my oldest friend Pat Kent, who had died of colon cancer some months before, showed up. He was clearly dead; his flesh had the wet and almost glowing aspect that signals the beginning of decomposition, but I was not revolted. I accepted this appearance as a sign of his being dead, since in another part of my mind I had not forgotten that he had been cremated. We had a conversation whose details did not persist when I awoke. But the effect was to let me know that Pat was still Pat, even though dead.

My recently-dead stepfather appeared in a similar dream, which also involved a conversation I cannot remember, and which also had the effect of confirming his real presence while not denying his being dead.

In the most recent such dream, another long-time friend, Bob Bourdette, appeared to me. Already at our first meeting thirty-five years ago, Bob was suffering from the diabetes that would eventually destroy his feet and attack his eyesight, his heart, and his whole constitution. But in the dream I saw a younger and completely healthy Bob, handsome, cheerful, and well-dressed as always. We talked, and I asked if he missed anything about being alive. His answer was an eloquent gesture toward himself, unmistakably meaning, "Look at me. What do you think?"

Outside of these dreams, I don't talk to people who've died, although I often find myself wishing they were around so I could tell them my thoughts. And there is a perfectly logical naturalistic explanation of these dreams. My unconscious mind is doing the reassuring here, letting me know that my memories of my dead ones are safe, though the objects of those memories are beyond the reach of consciousness.

***

Does the audience believe in apparitions from the dead when the ghost appears in *Hamlet*? Most of my contemporaries, I suspect, don't believe in ghosts, and they need to work at suspending disbelief when the ghost shows up. Renaissance audiences *may* have been more likely to believe in ghosts, but it didn't matter, because they were used to the convention of the murdered man's crying for vengeance.

"I am thy father's ghost," says the actor who's just risen through the trap door in the stage.

Shakespeare anticipates the audience's skepticism; that's why Horatio is there. Having already decided the ghost is fantasy, Horatio "will not let belief take hold of him." But the ghost convinces him, "harrows" him "with fear and wonder." Horatio sees the ghost first when Hamlet is not present, so that we in the audience don't get to think of the ghost as merely a creature of Hamlet's imagination. And then it turns out that Horatio is full of ghost lore — about the Roman graves giving up their dead before Caesar's assassination, about spirits having to return to their "confines" before dawn, and so on.

Hamlet is a play about communing with the dead, not only in the ghost scenes, but also in the graveyard scene near the play's end, where Hamlet becomes symbolically prepared to accept his father's death at last. Hamlet gets a short course in death, first in the form of anonymous bones flung up by the gravediggers, then more personally in the skull of the court jester Yorick, who used to carry young Hamlet around on his shoulders, and then, closer still, in the shrouded corpse of the woman Hamlet finally, unequivocally, admits he loved.

Community with these dead brings him past some of the horrors of death ("How long will a man lie i' th' earth ere he rot?"), past the unchecked emotion of first grief that he chastises in Laertes, and into a return to confidence in his own agency.

***

The ghosts in M. Night Shyamalan's movie, *The Sixth Sense*, are angry, like old Hamlet stalking the battlements. They want to be listened to, not just by anyone, but by a special person who can change things in the world of the living. But unlike Hamlet's father, they don't know they're dead, although those who have been killed seek redress.

The dead people in *The Sixth Sense* are not disembodied spirits but the recently dead bodies of those who still have the marks of their deaths upon them, in gunshot wounds or evidence of profound sickness or trauma. The bodies themselves seem to represent the reluctance of the spirit to depart.

It appears from these examples that the problem the living have with the dead, and part of the reason for wishing to converse with them about their situation, is the problem of the body. The horror of the decomposing body of one who is loved, along with the projection of our own body into such a situation, animates much of the literary and performance representations of the dead communicating with the living.

***

Walt Whitman attempts to depersonalize the relation with the dead in a poem called "This Compost". Hamlet's question, "How long will a man lie i' th' earth ere he rot?" is vividly performed in the poem, as the speaker asks "how can it be that the ground does not sicken" from all the "distemper'd corpses" that have been put into it. He addresses the earth directly:

> Where have you disposed of their carcasses...
> Where have you drawn off all the foul liquid and meat?

In the poem's resolution, the speaker marvels ("What chemistry!") at the earth's purifying compost, which ensures that "all is clean forever and forever," as the earth, which is "that calm and patient," transforms the dead and "grows such sweet things out of such corruptions." Whitman resolves the problem of the body in a very modern way: all of those dead, which he makes so impersonal but which his language reveals as very close in his imagination, are cleansed of rot by becoming one with nature. Although this consolation is as old as Epictetus, it fits very well with a modern sensibility that would rather see earth as a live thing than think of the spirits of dead relatives that way.

***

Stephen Greenblatt begins his book *Shakespearean Negotiations* with the words "I began with a desire to speak with the dead." Here, I think, is the essence of the historian's motivation. And do we not commune with the dead almost every time we open a book? For me, at least, reading is communing without conversation. But the historian's or the reader's connection with those who lived and wrote in the past is not what most people would mean by communing with the dead. Do they talk to you? Do you talk to them? I often hear, in my mind, one of my stepfather's lines summing up his experience ("I'd rather be lucky than good," he'd say as a golf shot that started badly ended up sweet). This doesn't count, I suppose. I speak to him or to dead friends, when something happens I know would amuse them. Again, probably not relevant. Nor, I suppose, are my dreams. Terry Barrett, a psychologist friend of mine, once told me that one out of every four of his experimental subjects had some kind of "auditory hallucinations", which in his definition could range from hearing one's name amid the noise of a busy street to having a full-fledged conversation with people whom other observers cannot see or hear. Anything one heard from the other side would necessarily be classed as hallucinatory by his working definition. But his number of twenty-five percent intrigues me. If fifty-one percent of the population heard voices from the beyond, would the forty-nine percent who didn't be justified in calling them hallucinations?

I know that I am leaving in the air the question of what constitutes *communing*. Where is it in the scale between indifference (chillier than the engagement represented by hatred) and complete merging with the other? Does the dance of slight misunderstandings we know as conversation make for more or less communion than mutual and wordless contemplation?

I do not make a claim to have held communication with the dead according to any strict definition of what "communing with the dead" might be, but I am not prepared to give up the possibility. I am more and more convinced that the desire to talk with the dead is so widespread as to be universal and that it is independent of religious dogma, which merely ritualizes or formalizes our behavior toward the dead, sometimes without touching our beliefs.

***

99

As I wander around Tzintzuntzan on the Day of the Dead, I see everywhere the marigold, called *zempazuchil*, a name that hearkens back to the flower's association with the Aztec princess, Xuchitl, whose name *means* flower. But though marigolds line the streets and pathways leading to the cemeteries, the flowers that actually adorn the graves are as likely to be *alcatraces* or Calla Lillies, *terciopelo* — the brilliant magenta cockscomb flower — and the tiny white blossoms of baby's breath, which the Mexicans call *nube* or cloud.

The odor of the flowers mixes with that of candle wax, and as evening turns into night, candles are everywhere, tens of thousands of them, sometimes hundreds at a single grave. They illuminate the offerings brought to the graves — food, drink, tobacco, toys in the case of dead children, photographs and sports gear, candy and sweets — as well as the faces of the vigil keepers. The night begins to cool, and the village lights in the distance go out one by one. The graveyards are now the village, filled with the living and the dead. A light mist forms in the lower spots, as the earth exhales.

10. A Dream of Learning

During the summer after my freshman year of college I sold expensive books door to door in Phoenix. The *Great Books of the Western World* — you can still see them occasionally for sale in used book stores — was a set of 54 volumes of classics in literature, history, philosophy, psychology, economics, politics, mathematics, and science, published by the same people who put out the *Encyclopaedia Britannica* and marketed as a way to acquire familiarity with everything worthwhile that had been written about "the great ideas" of Western civilization.

My college roommate and I answered an ad and spent two days learning the pitch for the Great Books. It was a canned speech that had to be memorized and then delivered word for word — no improvisation was permitted during the training period. Even afterward we were told our best chance for a sale was to stick closely to the text that a team of marketing psychologists had worked hard to get right.

"If you can get through the pitch five times", said Whitey, my gravel-voiced team leader, "you will get a sale." Whitey was about sixty, or at least that was the guess of a nineteen-year-old for whom old doesn't admit of too many degrees. He looked as if he would be equally at ease running a small bookie operation or grifting an insurance scam. The appeal of the pitch for him might have been its mendacity — not about the merit of the books, but about the way the customers were to pay for them. The pitch first offered ridiculously low payments over a ten-year period, but you couldn't close the deal unless you "converted" the buyer to a more expensive, 30-month plan. There was in fact no ten-year option.

But Whitey was right, at least in my experience. I hardly ever got through more than one pitch a night, working from shortly after five in the evening—we had to pitch husbands and wives together—until about nine-thirty or ten, when people simply stopped answering the door. I sold a set of books on average once every five days and made almost $500 for the month I sold books, which wasn't bad for a nineteen-year old in 1962 working a part-time evening job.

***

The Great Books of the Western World was not the first such experiment in popular education. Half a century earlier, two editors at Collier, the encyclopedia publisher, had talked the recently retired head of Harvard into putting together the *Harvard Classics*. Charles William Eliot, who had run the university longer than any president before him, had claimed in speeches to working men (shades of John Ruskin!) that "a five-foot shelf would hold books enough to give in the course of years of reading a good substitute for a liberal education." Thus the *Harvard Classics*, also known as the *Harvard Five-Foot Shelf of Books*, was begun. The fifty volumes Eliot eventually chose were heavy on literature and also contained ancient and modern philosophy, economics (Adam Smith but not Karl Marx), science of the nineteenth century (Darwin, Lister, Faraday, Pasteur) and earlier (Harvey, Jenner, but no Copernicus, Galileo, or Newton), politics, history, travels, and religious texts.

Eliot described his intent in the first paragraph of his introduction to the set:

My purpose in selecting *The Harvard Classics* was to provide the literary materials from which a careful and persistent reader might gain a fair view of the progress of man observing, recording, inventing, and imagining from the earliest historical times to the close of the nineteenth century. Within the limits of fifty volumes…I was to provide the means of obtaining such a knowledge of ancient and modern literature as seems essential to the twentieth century idea of a cultivated man. The best acquisition of a cultivated man is a liberal frame of mind or way of thinking; but there must be added to that possession acquaintance with the prodigious store of recorded discoveries, experiences, and reflections which humanity in its intermittent and irregular progress from barbarism to civilization has acquired and laid up. From that store I proposed to make such a selection as any intellectually ambitious American family might use to advantage, even if their early opportunities of education had been scanty.

***

I ring the doorbell of the newly-built tract house in the Phoenix suburbs.

"Hi, I'm Mike Cohen, University of Chicago," I say without a blush — another bit of mendacity in the pitch. "I'm here to talk to you about a unique educational opportunity. You're interested in education, aren't you, Mr. ------?"

The slight, blond man who has come to the door looks no older than I am. With only a brief hesitation, he gives me his last name, which means I can now use it in the pitch. He's just home from work and hasn't had time to change — a white, short-sleeved dress shirt, with a collar too large for his skinny neck, and a tie.

We sit down and I stall while his wife is putting the baby down in the next room. When she comes in I start the pitch in earnest. She looks even younger than her husband, also slight, and tired. They are sitting on the couch and I on a straight-back chair that is the only other furniture in the living room. Over their heads, through the undraperied picture window, I can see the bare yard with a sprinkler going on dirt that must be newly sown with grass. They haven't been in the house long enough for it to sprout.

"Start *there*," Whitey had said as he let me out of the team car, indicating the house with the bare yard. It didn't look very promising to me. But she was right. An hour later I left with my first sale.

***

The "ambitious American family" whose "early opportunities of education had been scanty" is the main target for the *Great Books* just as it was for the *Harvard Classics*. Of course they appealed to other buyers. Once I made a sale to a man who was clearly familiar with the names on the spines; he wanted the books as a resource but was also eager at the chance to just dip into authors he'd never read before. The extreme case was what Whitey called a "book mooch". This kind of customer was sold early in the pitch when the pitchmaker unfolded a full-size color photograph of the books — printed on a canvas broadside — and spread it at his feet.

"You can see the pupils of their eyes get bigger," Whitey said. "It's like they were watching porn or something."

The mooches tended to answer magazine ads featuring the *Great Books*, mailing in their addresses to have further information "sent" to them. These leads were always kept by Whitey and the other team chiefs and resulted in a much higher rate of sale than the cold canvas we newbies were limited to.

***

The Great Books took up slightly more shelf space than the *Harvard Classics*, and came with their own bookcase. Eliot had envisioned the earlier set as a series of courses; the *Great Books* were inspired by a course philosopher Mortimer Adler and university president Robert Hutchins developed at the University of Chicago around some landmark texts of the Western tradition. One of the students in the Adler-Hutchins course was William Benton, who, when he became publisher of *Encyclopaedia Britannica* in the 1940s, enlisted his former teachers in a project to select texts for what would become the *Great Books of the Western World*. By the time the set was published in 1952, the endeavor's educational purpose had acquired almost sacred status in Hutchins' view.

"This is more than a set of books," Hutchins told the crowd at the publication launch. "*Great Books of the Western World* is an act of piety. Here are the sources of our being. Here is our heritage. This is the West. This is its meaning for mankind."

Adler and Hutchins benefited from some of the criticism that had been directed toward the *Harvard Classics* over forty years. All of Homer, the Greek playwrights, Plato and Aristotle (two volumes) were included. So were many more science and mathematics texts. The English literature representation began with Chaucer (*Troilus* and *The Canterbury Tales*), whom Eliot had excluded. Not only Darwin, but Hegel, Marx, and Freud were amply presented.

The great novelty, however, was the addition of what Mortimer Adler called the Syntopicon. He and almost a hundred readers working for him made a list of ideas which they found recurring in the authors of the *Great Books*. They then painstakingly compiled an index of passages where each idea was discussed. References to 3,000 different topics were assembled, and then Adler reduced these topics to a master list of larger ideas. He called these the "Great Ideas," and they included such obvious choices as Beauty, Chance, Democracy, Fate, God, Good and Evil, Happiness, Justice, Religion, Sin, Truth and Wisdom. Some less obvious choices were Angel, Definition, Habit, Oligarchy, and Prophecy. The Syntopicon gave the *Great Books* something of the nature of an encyclopedia that could be dipped into and sampled by the use of this master index.

***

The *Great Books* sold poorly during the first few years. Then *Encyclopaedia Britannica* started to market the books the same way they marketed their encyclopedia. In the late 1950s they began using the pitch Britannica's marketing psychologists had painstakingly assembled; in 1961, the year before I was going door to door, Britannica sold 50,000 sets.

Selling the *Great Books* like an encyclopedia seems to be an inevitable result of their conception. Clearly the *Harvard Classics* and the *Great Books* both began in the minds of people who made encyclopedias; they were conceived for the common reader and the common reader was their obvious market. The encyclopedia is also the product of a dream of learning for the common folk, born of the Enlightenment, but housed in a profit-making form.

Encyclopedias of the non-virtual sort, made up of a row of sizable and very actual books, are rare now. In its inception though, the encyclopedia is remarkably like the current virtual variety on the Wikipedia model that is the work of a community of scholars and intended for the use of the ordinary citizen, anyone who can access it, the nonspecialist, the inquirer who wants quick and concise but also accurate and reliable information. Learning served up in easy helpings.

The first modern encyclopedia was the *Encyclopédie, ou dictionnaire raisonné des sciences, des arts et des métiers* (1751-1765) edited and partly written by Denis Diderot and Jean le Rond d'Alembert. The *Encyclopédie* was twenty-eight volumes of information and illustrations put together by a score of learned writers that included Voltaire, Rousseau and Montesquieu. It was a populist effort that included descriptions of ordinary trades (*métiers*) by the workers themselves, and its writers rejected the authority of church and state to make any claims of knowledge. Nothing less than "human knowledge in a truly unified system" (d'Alembert's words) was the aim of the *Encyclopédie*, and its intended audience was any one who desired knowledge and could read. Diderot and d'Alembert did not appeal directly to "any intellectually ambitious... family... even if their early opportunities of education had been scanty", but they did ultimately find such readers, and most scholars agree that the *Encyclopédie* contributed significantly to the democratizing movements leading up to the revolution in France. Of course it is the overall populist effect of the *Encyclopédie* through time that I speak of. The first sets were sold to rich men by subscription before they were even written, and there was never anyone going door to door saying, "Salut! Je suis Michel Cohen, de la Société des Philosophes."

Collecting writings such as the *Harvard Classics* or the *Great Books* is a less subversive activity than that of the encyclopedists. There are plenty of dangerous ideas in the *Great Books*, and the actual words of writers such as Locke show up in revolutionary documents like the Declaration of Independence and the French Declaration of the Rights of Man and of the Citizen. But the sedition and the blasphemy and the rabble-rousing words were all written by those long dead, who can't be hauled off to jail, as Diderot was at one point.

***

Large editorial projects like assembling great books for mass sale are a thing of the past. Too many of the works in such collections are in the public domain, and are therefore available free online as e-texts. But the lack of commercial motive is not the only reason why we don't see a twenty-first-century five-foot bookshelf.

Civilization's necessary documents probably need a much larger shelf than five feet. Partly this is a matter of expanding the notion of "western" to include texts from farther afield, but editors would also be thinking about women writers and scientists who never even occurred to Eliot (a few of these found their way into the second edition of the *Great Books*), and perhaps finding slave narratives and captivity narratives that would fill some of the gaps between Eliot's all-white writers. "And so on, endlessly, until we have inflated the five-foot shelf to the size of the Widener Library," writes Adam Kirsch in "The Five-Foot Shelf Reconsidered" in *Harvard Magazine* in 2001. But even if we could keep the size under control, the selections themselves would be less readable, because an explosion of knowledge has been accompanied by an acceleration in its specialization.

A more serious obstacle is the growth of skeptical attitudes. We are less inclined than we used to be to invest anyone with the cultural authority to tell us what the great books are. Strident voices are raised in criticism whenever anyone presumes to talk about the "best" books or the ones "necessary" to a liberal education. Moreover a contemporary skepticism retreats into the theoretical, into a kind of meta-specialization. We seem convinced that no one can be educated in history by historians without hearing from historiographers about the biases and blind spots of the historians — and so on in every other discipline.

Yet there are testimonials. David Denby writes about the joys of re-taking the Columbia University great books class he first took as a freshman in 1961. Sometimes it is hard to tell whether Denby's *Great Books: My Adventures with Homer, Rousseau, Woolf, and Other Indestructible Writers of the Western World* (1996) is a paean to the genuinely indestructible writers or an elegy for a kind of instruction that was common enough during the twentieth century, but is mostly gone or profoundly diluted in this one. Nonetheless, there are still many colleges with Great Books curricula, either as the multiyear core of liberal arts degrees or merely as a one- or two-semester survey.

And outside of colleges there are also still many listmakers. Publishers, newspapers, and magazines often come out with lists of the hundred novels we dare not die before we read, the vital nonfiction books of the last hundred years, and so on. Robert B. Downs' *Books That Changed the World* has gone through three editions since 1956 and remains in print. Martin Seymour-Smith's *The 100 Most Influential Books Ever Written* got a great deal of attention when it was published in 1998. Except for a score or so of writers, these lists rarely give us a consensus, and their more outré choices often tell more about the compiler's personality than the galaxy of important books. That there is no consensus does not preclude a census, of course. The catch is that you have to be educated to undertake an intelligent census, to find out on your own just what constitutes the best that has been thought and said. In fact, the older and better-stocked the mind, the more pleasure it takes in making choices about what to read.

***

One can still buy the *Great Books* either in an expensive second edition put together in 1990 or in the original edition, from used book dealers online or in brick and mortar stores. The sets are easy to find since more than a million were sold. The *Harvard Classics* also show up in bookstores, and in recent decades Easton Press produced a leather-bound set and Kessinger Publishing a paperback one. But the entire five-foot bookshelf is now available online from Gutenberg.org. Another site, MyHarvardClassics.com, has all the volumes available as pdf file downloads, and the site also has a reading plan: "Use this accelerated reading guide for 1 hour a day and become a cultivated scholar with all the elements of a liberal education in 90 Days." The modern hawkers of self-education are those who've made a market in homeschooling and in denying the necessity of college. UnCollege.org, sells its founder's book *Hacking Your Education* and various expensive "programs" rather than courses. The Great Books Academy sells its courses for about three thousand dollars each.

In America we dream that everything should be within our reach — and without reaching very far. Some of us remember when milk and ice and baked goods, fresh vegetables and fruit, ice cream and knife sharpening, pot and pan repair and household odd jobs all were there at our doorstep or passing in the street. I grew up selling various things door to door: doughnuts and newspapers, little cans of Cloverine Brand Salve to get my first Daisy BB rifle, and eventually the

Great Books. I have been hawking great books all my adult life, selling them door to door my freshman year and teaching them in college Honors Humanities classes for years before I retired.

Having tried it both ways, though, I confess there has always seemed to be something mendacious about the door-to-door sale of education, or its modern equivalent in the internet do-it-in-the-comfort-of-your-own-home salesmen. I don't say the autodidact is a modern impossibility, but educating oneself was never easy and has to be more difficult now. There are no shortcuts. You really do have to read Euclid (or to have been taught geometry) in order to read Newton, and although you might not need any preparation to read Darwin, you jolly well won't grasp much from Kant's *Critique of Pure Reason* unless you've read Aristotle and Plato as well as Descartes and Hume, or had at least one good introductory philosophy course.

The dream of learning is in everyone's grasp, the dream of the enlightenment of a free and self-educated citizenry, has always been there, but in America it has always smelled a little too much like snake oil. Whitey didn't say this in so many words, but I think I knew it, even at nineteen.

11. Notebooks

My tables — meet it is I set it down.

> – *Hamlet*, 1.5.108 (c. 1600)

Hamlet carries with him his "tables", or tablets, notebooks in which young Renaissance gentlemen copied out quotes and passages that they found pithy or well-put. We see Hamlet writing in his at one point in his play. He's just been visited by his father's ghost. The ghost tells of his murder by his brother, Hamlet's uncle, who also seduced Hamlet's mother. The ghost demands revenge. Hamlet first talks about his own memory as a kind of notebook or "table". He says he will erase "from the table of my memory" everything except his father's command. Then he gets out his actual notebook:

> My tables — meet it is I set it down,
> That one may smile, and smile, and be a villain.
> At least I am sure it may be so in Denmark.
> So, uncle, there you are.

The revelation Hamlet has just received from the spirit world — his father murdered, his mother seduced, his uncle the seducer and the murderer — these all get reduced to a *sententia*: "one may smile, and smile, and be a villain." Somehow, between the revelation and its setting-down there occurs a reduction, so that to get it into the two dimensions of the little book it becomes smaller and, in every way, flatter. Could it be that it is in a notebook's nature to diminish experience? And if so, how could we hope to do better than Hamlet, the

Prince of Words, in this regard?

We would like to pretend that our notebooks are there for recording observations, other people's memorable words, things that we have found in our reading that "oft were thought but ne'er so well expressed." Our model here is one of Joyce's notebooks that he used while writing *Ulysses*, notebook VIII.A.5, completely filled with notes on Joyce's reading about the *Odyssey*, Homer, and Greek mythology, without interpretation or elaboration. But other Joyce notebooks, especially those he used for *Finnegans Wake*, are almost completely filled with his own thoughts.

Oliver Wendell Holmes tells us that the first two papers he published under the title *The Autocrat of the Breakfast Table*, written twenty-five years before the *Atlantic Monthly* columns eventually collected under that title, weren't worth reprinting. Holmes does tell us a couple of things that were in those two papers, published in *New England Magazine* in 1831 and 1832. One of them is a single sentence: "It is a capital plan to carry a tablet with you, and, when you find yourself felicitous, take notes of your own conversation."

We don't want to admit this real purpose of the writer's pocket notebook. It's all about *us*, rather than about our keen reporter's sense of observation. "For the past ten years or so," David Sedaris writes in one of his funny stories for *The New Yorker*, "I've made it a habit to carry a small notebook in my front pocket. The last page is always reserved for phone numbers, and the second to last I use for gift ideas. These are not things I might give to other people, but things they might give to me..."

I think the most candid approach I've heard of was that of Stanley Fish; in this case his notebook was a tape recorder. My friend Don Hedrick tells the story of a summer NEH seminar on cutting-edge criticism led by Stanley Fish and held at Johns Hopkins. Fish brought to the seminar room a small tape recorder — one of the modern permutations of the notebook — and turned it on as he began the day's proceedings. When the students picked up the discussion, he would turn the tape recorder off. When he himself weighed in, he would turn it back on. One never knows, until one plays the tape back, whether one will find oneself felicitous or not. Had the *Autocrat of the Breakfast Table* been a world-famous university professor like Fish, one can imagine him saying to his students, "If we want a seminar in *your* opinions, we'll schedule one."

Not that the notebook lacks its utilitarian aspect. My drawing teacher used to point to his pocket calendar-diary, stuffed with memos, pink "While You Were Away from Your Desk" telephone messages, and yellow Post-It notes, and say, "This is my brain." He was young at the time, and probably could have remembered appointments and meetings. But substituting the notebook for the table of his memory relieved him of having to remember. "For lack of a natural memory I make one of paper," writes Montaigne. My wife Katharine has at least a dozen little notebooks (spiral-bound so she can push the pen or pencil in the back) scattered around the house and her car to keep track of birds seen that day or season, rare license plates spotted while driving, calories or Weight Watcher points, and the things that need doing that day.

If you type "notebook" into an internet search engine, you will be led to sites featuring laptop computers rather than bound paper products. If you ask for a notebook at the store, you may be shown ring binders or small spiral-bound pads of paper, because what you want is often not called a notebook these days, but a "journal". For $32 you can buy a journal with a handmade cover, acid-free paper, gilt page edges, and a

satin ribbon page-marker. You can buy a Harry Potter
journal, a gardener's journal, a book club journal with
suggestions about books for your club to read, journals
with white or cream or psychedelic-colored pages that
are lined or plain or graph-ruled with cloth, leather, or
paper covers. Barnes & Noble has no fewer than seventy 3-foot shelves devoted to notebooks and journals.
One set of journals reproduces the covers of Nancy
Drew mysteries; another, famous paintings of women
reading. This move from notebook to journal may be
an aspect of the modern cult of the personality, another
sign that what the notebook is destined to record is just
us. We have met the notebook, and it is us.

And notebooks have entered an odd zone of ro-
mance. The trend may be exemplified by the Italian
brand known as Moleskine (both *es* are sounded). These
have sewn pages and are bound in cardboard which
has a textured oilcloth cover that could be imagined
to resemble moleskin. The notebooks are small — three
and a half by five and a half inches — and feature an
attached elastic band to keep the notebook closed and
a gusseted pocket in the back to hold loose papers. This
style of notebook, if not the current brand, was used by
Bruce Chatwin, Ernest Hemingway, Pablo Picasso, and
other art luminaries, if we are to believe "The History of a Legendary Notebook", printed in six languag-
es and packaged with each Moleskine. I found several
websites devoted to how various contemporary writers and artists, none of whose names were familiar to
me, have filled their Moleskines, and you can play the
game of identifying Moleskine notebooks in the hands
of actors in various movies and television programs.

Lauren Hutton, looking pretty good after a model-
ing career followed by an acting career, waves a little
journal around during her now notorious Wyeth commercial for hormone replacement therapy and is seen

writing in it, though we don't get to see what pearls she has recorded. Writing in her journal is represented as part of the good life of a strong, attractive, not-so-young woman. It looks like it might be a Moleskine.

The type David Sedaris favors is called The Europa — according to its website, a spiral bound, pressed-board covered notebook three by five inches containing 60 ruled leaves. One series is available in dark blue, red, green, yellow, or lilac; the other in lime, lemon, black, pink, or fuchsia. The notebooks, which are made in the UK, will set you back about a pound apiece if bought in quantity. And then there's The Rhodia, "the French orange notebooks with a cult following", as one online dealer describes them.

At Duke University's English and Literature Department, I am informed by a reliable source, when famous lecturers come to talk, around the seminar table or the lecture room some very fancy notebooks and expensive writing instruments come out of pockets, briefcases, and purses. Perhaps a competition (I think of the scene in *American Psycho* where the modish young executives try to one-up each other with their gilt or color-engraved business cards on hand-laid papers), perhaps merely necessary status-marking, the scene contains some ironies: many of the people in the room declare themselves Marxists and in their writings deplore the fetishization of commodities and objects.

My original pocket notebook was graph paper. I bought it in a London bookstore. It had plastic-treated covers and I remember being amazed at how long it stayed together — in fact, I came home from the trip and used it for weeks afterward.

When I was traveling overseas for six or eight weeks at a time, I needed a notebook small enough to put in a pocket, with enough space for a journal and also for planning notes, copying addresses and telephone numbers, making lists of gifts to bring back (for others, not for myself), making notes on my reading, and jotting occasional thoughts.

I tried various kinds of notebooks over the years. I had a brief experiment with the Moleskine ones with stiff cardboard covers. They are nice notebooks, with an elastic band to keep them closed, a sewn-in ribbon placeholder, and a little pocket inside the back cover to hold receipts and tickets. But they have more than a hundred pages with stiff covers, so when I put them in my back pocket I found myself sitting at an angle, as if I were always sailing to windward. The Moleskine was too bulky for a front pocket, and as an academic — therefore summer-traveler I usually wore no coat to put it in. The thinner variety Moleskine notebooks, called "cahiers", looked like the ideal answer at first. These have only 64 pages (32 sheets, with the last 16 perforated so they can be removed), with a more flexible cardboard cover and a pocket in the back. The detachable pages sound better than they actually prove to be: I don't rip out a note from my notebook that often, and the perforated pages can begin to detach themselves after several weeks of hard use. The pocket in the back is glued at the bottom and not gusseted like its prototype, so it's not very useful.

I thought I'd found the ideal notebook in the 50-sheet "Li'l Comp" made by Mead. Cheap, flat enough to fit in my back pocket and strong enough to keep from disintegrating (the pages were sewn rather than stapled), it was perfect for a month of jottings. Then Mead stopped making it. They wrote me a consoling letter: "We apologize for any inconvenience this may cause you." I began to appreciate the brand loyalty of buyers of the fancier brands of notebooks. Walt Whitman liked small, leather-covered notebooks, and when he could not get the kind he wanted, he resorted to making them himself.

What I need, no one makes: a thin notebook that will fit in a front pocket, at least enough pages for a month's daily notes, with a calendar for the month, spread across two pages with a block for each day big enough to write in, with sewn pages so it doesn't come apart, and soft but strong covers. I end up pasting a calendar in whatever notebook will fit the size requirement. I need only a month's worth of pages and a month's calendar because I transcribe daily notes into a computer file; it's tedious, but it's the only way to make the material useful for searches.

I lack the ingenuity of Walt Whitman, who not only made his own notebooks sometimes but almost always modified the ones he bought. Over a hundred of his small notebooks, with lecture notes, comments about his reading, drafts of poems, and general observations, have survived. These often have pages cut out, new ones pasted in, clippings, photos, and other additions. But mine do have monthly calendars pasted or drawn in, and each month ends with my notebook stuffed with receipts, tickets, and other additions.

Sometimes notebooks turn into journals. In *Innocents Abroad,* Twain talks about thirty or so of his fellow passengers on the boat across to Europe starting journals, speculating that not one would be still writing at the end. "Alas! That journals so voluminously begun should come to so lame and impotent a conclusion as most of them did!" He says the impulse is universal. "At certain periods, it becomes the dearest ambition of a man to keep a faithful record of his performances in a book," but that man has to have "pluck, endurance, devotion to duty for duty's sake, and invincible determination" to keep at it. "If you wish to inflict a heartless and malignant punishment upon a young person, pledge him to keep a journal a year."

I kept a sporadic journal during the year I took

off from college to travel — my twenty-first year. But I didn't try again for twenty years until a three-week trip to Spain in 1984. That trip started a travel journal that I kept up whenever I was on the road, whether for a weekend conference or a several-months sojourn in another city.

After twenty more years, I now find myself keeping not one but several journals. I have a birding journal that records new birds, together with the place and date I first saw them, and occasionally lists of birds on productive days or on trips taken specifically for birding. I keep an astronomy journal that records celestial sights with dates, times, and locations I saw them from; I also record seeing conditions, telescopes, eyepieces and filters used, and brief descriptions of the objects. I keep a pilot's log every time I fly. And my travel journal has turned into a record of non-travel events as well. I don't necessarily record something every day, but I certainly do every week. The problem of keeping a journal for me has less to do with endurance and determination and more to do with banality, with how to keep from its becoming "Went here; went there; ate this; did that." Of course, that kind of record has its uses, and in those initial days in Spain, where I went and what I ate were often spectacular. But the everyday soon overwhelms the spectacular, and even Hamlet's world-shaking realizations reduce to "One may smile, and smile, and be a villain."

The problem is still the same: the notebook seems to make one's experience smaller. And yet, Joyce and Whitman never were bothered by any such diminution. And I remember seeing, in 1980, one of Turner's notebooks open in a display case at an art gallery in York. Turner used over 300 notebooks during his career, primarily for sketches but also for travel notes, casual thoughts, personal finances, and descriptions of the weather. This particular notebook was tiny — about two by three inches — but had sketches of sweeping landscape. The whole atmosphere on a postage stamp. Ah, a *sententia*! Meet it is I set it down. Now, where is my notebook?

12. A Fountain Pen of Good Repute

Writers of all sorts fetishize their tools. Hannah Arendt apologized to Martin Heidegger for using a typewriter for her letter (9 February 1950); "because my fountain pen is broken and my handwriting has become illegible." Roland Barthes wrote everything by hand, using different fountain pens from his collection, often switching from one pen to another, as he told an interviewer, "just for the pleasure of it."

A more practical inclination is that expressed by Joseph Conrad in a letter to his agent, J. B. Pinker (21 February 1906), begging for a good fountain pen. Conrad sends part of what would become *The Secret Agent* and adds this postscript:

> PS Would you have the extreme kindness to buy for me
> and send out [Conrad was in Montpellier, in the south of
> France] by parcel post a fountain pen of good repute —
> even if it has to cost 10/6. I am doing much of my writing
> in the gardens of Peyron under a sunny wall and the horri-
> ble stylo I've got with me is a nuisance.

The *stylographic* pen to which Conrad refers was an early forerunner of the ballpoint: a cylinder of ink had a hole at the bottom closed with a pointed metal pin. Pressure on the pen pushed the pin up into the barrel, releasing ink. When the pressure eased, the pin resealed the hole. A stylographic pen is also mentioned in Arthur Conan Doyle's *The Lost World* (1912).

Beyond aesthetics and convenience comes an identification of writer with pen: the feeling that the pen itself is somehow doing some of the creative work of the person who wields it, or that the pen is an extension of the self and actually gleans thoughts from the teeming brain faster than the brain can provide them. "My two fingers on a typewriter have never connected with my brain," Graham Greene said. "My hand on a pen does. A fountain pen, of course. Ball-point pens are only good for filling out forms on a plane." There are no doubt analogies in the way wooden-boat builders feel about their hand planes or spoke shaves, or carpenters about their favorite hammers. "I am a man-pen," wrote Flaubert, "I feel through the pen, because of the pen."

The Montblanc fountain pen I hold in my hand—a retirement gift from my wife—bears little resemblance to the instruments people have used to get ink on parchment or paper before the modern pen was invented, yet functionally and structurally, it is remarkably like a quill or feather pen (the word *pen* derives from the Latin word *penna*, or feather; the augmentative form of the Latin word gives us *pinion*, or wing). Quill pens were ordinarily made from the large primary wing feathers of geese. The hollow shaft of these feathers was big enough to serve as an ink reservoir for several minutes of writing. After hardening the quills in hot water, ashes, or sand, the pen-maker cut the quill's tip diagonally from its bottom (the inner curve of the feather) toward its top. Then the writer shaped the nib for the breadth of stroke she liked and cut a quarter-inch slit in the nib to facilitate ink transfer and allow the nib to spread under writing pressure.

The lightness of touch that suffices to make lines flow from a fountain pen is difficult to believe for one used to pressing down on a ballpoint. A fountain pen requires no pressure other than its own weight to make ink flow. It will write acceptably with the cap removed, but placing the cap on the top of the pen improves the balance and slightly increases the weight bearing on the nib to give a better and more effortless feel. There is a technical term for placing the cap on the non-business end of the pen: it's called *posting*, and the *posted* cap seems a horsey metaphor: jockeys sitting ready on their mounts at the track, or times when letters written with such pens were given in at the post office to be carried by horsemen riding post.

When I look down at the top of the nib on the pen I am using, I see a shape very similar to that of a cut nib on a goose quill. My nib is made of metal of course, and it has a further refinement in shape: the slit in the nib ends in a small hole, called a "breather", that allows air to enter and take the place of the ink as it is exhausted.

The superiority of metal nibs was immediately apparent once the dip pen began to be made in the late eighteenth century. The permanence of the nib made up for the fact that the dip pen had no reservoir of ink and had to be dipped every few strokes. And the simplicity and durability of the metal-nibbed dip pen gave it tremendous longevity. These were the pens that had been used with those inkwells in the desks of my youth. The pens themselves were no longer in the schoolrooms where I sat, but I remember them in the long high tables at the Post Office, though I cannot seem to recall a single inkwell set into those tables that was not dry. The metal-nibbed dip pen still has advantages over the fountain pen in one respect: it can handle any kind of ink: India and other sorts of indelible inks made with pigment in suspension, acrylic inks used in illustration and cartooning, and the ink that for centuries has been made from iron dissolved in sulfuric acid and the tannin from tree growths called galls. The familiar faded brown writing in such documents as the Declaration of Independence comes from their iron gall ink, blue-black when first applied and fading to brown with time and exposure to light. These inks would fatally clog the works of a fountain pen, and in fact fountain pens work best with water-based inks that keep them writing smoothly but are necessarily not indelible. My mother-in-law, who has always written her letters and cards in ink, gets around the problem of addresses washing off envelopes in wet weather by covering them with transparent tape. Lately new waterproof inks that use paper-bonding dyes rather than pigments are becoming available for fountain pens.

Steel dip pens do corrode over time, and they tend to be scratchy. The modern fountain pen has been rendered smooth-writing by the use of noble metals for the nib, which ordinarily is some alloy of gold, while the very tip of the nib may have a coating of harder stuff such as rhodium, iridium, or platinum. Fountain pen nibs were once more flexible, but the advent of the ballpoint has brought into use a stiffer nib more useful for bearing down on carbons and other manifold copies.

But why a *fountain* pen? It's the last image we want as pen users—a fountain or gush of dark stain coming from the instrument we want to lay down its ink in neat lines. Nancy Mitford hopes her fountain pen will behave as she writes to Evelyn Waugh in 1945: "It has been more of a fountain than a pen lately."

Badly capped or not "bled" of excess air before an airplane flight, stored nib-down instead of nib-up, not sufficiently controlled or handled deftly enough while over the ink bottle—all such scenarios result in squid-squirts we may poetically, ruefully call fountains. In fact, the pen historian Glen Bowen tells the story that Lewis Waterman, the man who made the fountain pen a commercial success, was inspired to perfect it because of an early model pen he'd purchased that embarrassed him when a potential client tried to use it and ended up with nothing but puddles of ink on the application for insurance Waterman was selling.

The word *fountain* may have been applied in inventories to a pen as early as the seventeenth century, but the first time we can connect the word to an artifact is in a British patent application from the 1820s. By comparison to the old method of dipping the pen every few words, writing with a more or less continuous flow of ink must have seemed to the pen wielder as if the ink was coming from a fountain. But the analogy of the ink flowing like water from a secret place, aided by gravity and capillary action, seems more like an artesian well or spring than a fountain. Calling it a *reservoir* pen, as some pen scholars do, is a little pedantic, so fountain pen it was and is.

The mechanism of fountain pens grew in complexity and ingenuity during the hundred years of their heyday from the middle of the nineteenth to the middle of the twentieth century. The first such pens had a reservoir filled with ink from the top, with an eyedropper. Later came a rubber tube—the eyedropper internalized—that could be compressed, expelling air, and when the pressure was released it would suck ink into itself. The compression was accomplished with a small lever set into the pen barrel in the very popular lever-filling system introduced by Walter Sheaffer at the beginning of the twentieth century. The lever system had great longevity in pen making, and most of the fountain pens I remember from the 1940s and 1950s were of this sort. But there were dozens of designs for filling pens. A capillary system of small tubes filled itself when dipped in ink, but tended to get incurably clogged. The piston filler, probably the most common one in pens manufactured now, came into widespread use in the Pelikan pens of the 1930s. A cap at the top of the filler turns a screw mechanism that raises and lowers a piston inside the barrel of the pen, expelling air on the downstroke, pulling up ink on the way up. The most ingenious of fillers was probably the Sheaffer

Snorkel, introduced in the early 1950s when fountain pen manufacturers were already feeling their markets disappear. The snorkel had a small retractable tube under the nib that went into the ink so that the whole nib did not have to be submerged; it was a much cleaner process, and it was introduced, I remember, with all the ad fanfare of a new car model.

When Anne Frank was nine years old, in February, 1939, her grandmother sent her a fountain pen as a gift. We don't know what kind of pen it was, though some modern collectors believe it to have been a Montblanc from her sketchy description. We know it had a gold nib, that it came in a red leather case, and that she treasured it. "My fountain pen has always been one of my most priceless possessions," she begins her account of its loss.

When in June, 1942 she received for her birthday a blank book with a lock, she began to keep a diary, writing entries every few days with her favorite pen. Pen and book went with her three weeks later into the "secret annex" in the back of her father's office building in Amsterdam, where her family went into hiding from the German occupiers after Anne's sister Margot received a notice to report to a work camp.

In November, 1943, Anne, now on her second diary, put her pen down on a table and was about to start writing when the rest of family interrupted her. Beans were spread on the table to be cleaned. When Anne finished helping with this task, she swept the floor and put the discarded bean husks into the stove in the room.

Afterwards Anne could not find her fountain pen. When Margot suggested it might have been accidentally thrown into the fire, Anne refused to believe it. But the next day her father found in the stove's ashes the pocket clip of Anne's pen, with no trace of the celluloid barrel or gold nib. "One consolation remains to me, though a slender one," Anne wrote in her diary. "My pen is cremated, which I would like to be later." Then she adds a little doggerel verse: "The clip is left, / (The pen is gone) / I feel bereft / And put upon."

Anne's father, Otto Frank, the only family member to survive the concentration camps, published her diary after the war. He omitted a number of passages, and here he chose to leave out the little doggerel quatrain. Perhaps he thought it was too flippant. More likely he wanted that last sentence about cremation to get the emphasis that the ending gives. Just possibly he recognized it was itself a verse. Notice, in the original Dutch, the metric parallelism of clauses (despite the extra syllables in the last one) and the near rhyme:

> Eén troost is mij gebleven, al is hij maar schraal:
> mijn vulpen is gecremeerd, net wat ik later zo graag wil!

The story of Anne Frank and her pen does not end there. Robert Faurisson and other Holocaust deniers have attacked her diary as a fake. Part of the argument has been the assertion that the diary was written in ballpoint, an anachronism, and therefore must have been forged after the war. The Netherlands Forensic Institute, at the request of the Netherlands Institute for War Documentation, studied the diary manuscripts and issued a report in 1986. Their conclusion was that two loose scraps of paper inserted in the diary pages were in ballpoint, as were page numbers added to the manuscripts after they were written. All the rest of the pages, the Institute concluded, were written by a fountain pen in the same handwriting as known samples of Anne Frank's writing. Every ink trace of the diary thus acts like a signature, insuring that Anne Frank's identification with her pen will remain indelible.

The fountain pen does not belong in some environments. It won't write in the zero gravity of space, and can leak in the low-pressure environment of airplane cockpits and cabins. But the point here is partly that the pilot doesn't really want to be bothered with thinking about writing with this or that instrument; her business is flying the plane, a thought that leads naturally on to the question whether the fountain pen belongs to any situations other than those in which writing *is* the business.

Fountain pens are obsolete. They have been superseded by the ballpoint and its variations. Of course a conventional ballpoint won't write in space either, and hence we have the further development of a pressurized ink cartridge with a high-tech ink, called thixatropic, that remains semi-solid until subjected to the shearing force of the rolling ball. This pen, invented by Paul Fisher — not *for* the space program, as is commonly thought, but adopted by NASA later — will write in extreme cold or heat, upside down or in zero gravity, and underwater. These are not realms in which the fountain pen can compete.

Fountain pens now function partly as commodities for collectors and status-seekers. John Windsor, writing in *The Observer* in 2007, traces three waves of interest in fountain pens. During the 1920s, before these pens had competition from ballpoints, "manufacturers first developed the coloured plastics that turned the fountain pen into the fetish object it still is today," writes Windsor. Then in the 1980s, "Thatcherism, the fashion for 'power accessories' and the increasing popularity of collecting as a hobby put a premium on classic pens." And about the current interest in fountain pens, Windsor says people have started buying them "for the strangest of reasons — because they want to write with them."

History will not be reversed. The pen will not be rendering any other writing device (or writing, revising, transposing, spell-checking, cutting-and-pasting, storing, organizing, and retrieving device) obsolete by its return. But fountain pens will not be going away soon, either. For those who use them, they seem an extension of the self, with the mixed feelings other parts of the self bring. "None of us can have as many virtues as the fountain-pen," wrote Mark Twain, and then he added, "or half its cussedness; but we can try."

13. Flying Lessons

Taking off in any airplane has its thrill. "Few seconds in life are more releasing than those in which a plane ascends to the sky," writes Alain de Botton in *The Art of Travel*. But a small plane delivers the feeling more satisfactorily. An airline passenger sees a tiny picture of the outside world framed by the plastic walls of his container. The cabin of my Cessna 150, on the other hand, has windows on all four sides, providing a panoramic view out and below.

When I take off, the whole point of this expansive greenhouse of windows asserts itself. The earth right under me drops down while the horizon shoots outward, remaining at the same apparent level, so that the effect is not that the whole earth recedes, but that it becomes an ever larger, shallow bowl. Space expands; as Samuel Hynes says: "The world is enormous: the size of the earth increases around me, and so does the size of the air; space expands, is a tall dome filled with a pale clean light, into which we are climbing." In a small plane, taking off is all light and surrounding space and a feeling of being way up in the middle of the air. And that moment when the jolt and vibration of wheels on the ground suddenly disappears as the plane is airborne, the unmistakable and exhilarating moment of moving from two into three dimensions is, as Isak Dinesen wrote, "the joy and glory of the flyer."

The sensation of an expanding world continues once I am aloft. Much more is visible from the air than from the ground, of course: on clear days I can see thirty-five miles by the time the plane has climbed to a thousand feet; fifty miles by two thousand, and so on until I can see a hundred miles. I also see differently. Hynes suggests it is a kind of voyeurism: nothing can hide itself from the aerial view, but the people on the ground are unaware that the little plane up there can see so much.

The earth looks more regular from the air. One of my passengers said, "Everything seems more organized somehow than it does on the ground," and I pointed out to her that even junkyards from the air have a colorful symmetry and even beauty. "No land is ugly from the air," writes Hynes. William Langewiesche goes so far as to say that "the greatest gift" of being able to fly is "to let us look around". But he does not insist that the view is beautiful — rather it is honest. Almost always it shows how much humans have put their mark on the landscape. Flying gives us a new view of our surroundings, even — perhaps especially — those that are most familiar to us.

2.

When I was in my teens my stepfather owned with another doctor a half interest in a Cessna 175, a four-seat single-engine airplane that we flew on trips to the coolness of northern Arizona, to visit relatives' in California (we flew into the tiny airport at Tustin, no longer tiny and now called John Wayne Airport), to Las Vegas, and in one long trip to Nassau, with stops in Texas, New Orleans, and Fort Lauderdale. But in the years between my teens and my retirement at sixty, I had not been in a small plane more than two or three times. That was one of my two mistakes that had to do with flying. The second one had to do with Seth.

My stepfather died in 2002, and a few months later I put his ashes in the car and drove over Gates Pass in the Tucson Mountains to Ryan Air Field. There I was going to take Seth's ashes up in a plane and scatter them over the mountains as he'd wished. At a flight school I was introduced to a pilot named Regan, who looked to be in his early twenties, and we went into the hangar. Regan took Seth's ashes out of the plastic bag and box they were in and put them in a brown Safeway grocery bag with little paper handles. He tied an eight-foot length of light rope around the top of the bag. Then we wheeled out the airplane, a Cessna 150 with two seats, and Regan and I got in. It was a tight fit, and I remember thinking I wouldn't want to log a really long cross-country in this one. As we took off, I once again had that feeling of heady release that I remembered from childhood experiences with flying. It occurred to me that this was Seth's last flight. It was, in some senses, my first. Once we got a few miles from the airport Regan said "Why don't you fly it?" I put my feet on the rudder pedals, took the yoke in my left hand as he directed, and did a few tentative banking turns. Then I tried to keep the plane going straight and level.

We flew over the Duval Mine, which I'd never seen from the air before, and toward the Santa Rita Mountains. I had wanted to scatter Seth's ashes in the Santa Catalinas, but since early June the Catalinas had been on fire. I decided on the Santa Ritas, which Seth had looked at from his back yard every day during the last years of his life. The smoke stretched halfway down the valley toward us now, but we were flying away from it. Once we got across the wash and into Madera Canyon, Regan opened the door and let out the length of rope with the bag at the end. Had he tried to let the ashes go in the doorway, they would have been sucked back into the cockpit. This way, the airspeed and prop wash ripped the paper bag apart in a few seconds. Then I did a hundred-eighty degree turn out of the canyon and flew back toward Ryan field. We practiced the landing maneuver — throttling back, lowering flaps, throttling back some more, keeping the nose level or slightly elevated to bleed off airspeed. When we got to the runway Regan said "Just fly down right above the ground and pull out the throttle, let it land" — and I did.

I had a quiet lunch at one of Seth's favorite restaurants in Tucson. I wasn't sad; I was exhilarated. I was going to learn to fly. I was sixty years old. What had taken me so long?

3.

The instructor who got me through my training to solo, and ultimately to my private license, ran an FBO at the airport in Murray — a Fixed-Base Operator provides flight services such as instruction, plane rental, fuel, and mechanical maintenance. Ernie had worked as a machinist in the local small-engine factory and retired to his dream of teaching flying and doing airframe and powerplant maintenance on the small airplanes in and around Murray. He is a year or two younger than I, patient, easygoing, and a perfect fit with my learning style. I get concepts quickly; putting them into coordinated action takes me longer to learn. Even now, many hundreds of hours after my training, I often hear Ernie's voice next to me in the cockpit. When I level out after takeoff I hear him say "Let the airspeed build up before you throttle back" and turning onto the base leg when I'm landing (we fly a path describing a rectangle; the base leg is the one just before turning to line up with the runway) his voice in my ear says "Keep the nose down and the airspeed up."

The voice in the cockpit was especially important during solo. This odd rite of passage puts you alone in the airplane for several takeoffs and landings, suggesting you know how to fly it, when in fact you know very little more than how to take off and land on a calm and uneventful day. When I soloed, I was surprised, as everyone is, at how lightly the airplane lifted off the runway with no weight in the right seat, but Ernie's voice was still there in my right ear. "Lift the nose as you reach sixty... now level off and build up airspeed... climb out at eight... wings level... stay over the runway centerline... standard bank onto crosswind leg... radio that you're staying in the pattern... tell 'em you're on downwind... throttle back to cruise as you reach pattern altitude... abeam of the numbers pull the carburetor heat on and throttle back to sixteen hundred... lift the nose to bleed off airspeed... ten degrees of flaps... tell 'em you're turning base leg... another notch of flaps... ease the throttle back on final... flare... hold the nose off... hold it off..."

The airport in Murray, Kentucky sits at 577 feet above mean sea level. Relatively flat landscape here stretches from the plains beyond the Mississippi eastward to central Kentucky and down toward the Gulf of Mexico. Western Kentucky is in many ways an ideal place to learn to fly. Aside from some tall towers near Nashville and Cape Girardeau, nothing prevents flying at 1500 feet across this terrain for hundreds of miles — a great advantage in a little plane that may take fifteen minutes (and a lot of fuel) to climb to an altitude sufficient to negotiate even small mountains. The small farm plots tend to be laid out according to the cardinal points of the compass, so that when the instructor says "make a right turn to 270°", even the novice, if she has any idea which way is north or south, can make the turn and stop it without even looking at the compass. And when she does look at the compass, it will be spot on. The agonic line, the corridor along which there is no variation between compass north and true north, runs very close to the Mississippi River here. By contrast, the line that runs through Tucson indicates an easterly variation of more than ten degrees in compass readings from true directions.

Within a hundred nautical miles of the airfield where I learned to fly are eighty public airports, and the nearness of a good landing surface makes for a comforting feeling for the nervous newbie. All but three or four of these airports are untowered, which means the newbie doesn't have to worry about a controller talking back so fast the student pilot has trouble following what's being said and controlling the airplane at the same time. There can be some strange radio effects on this flat land, by the way. The range of airplane radios lengthens when no obstacles interfere with the line of sight between transmitter and receiver; when I get up to a thousand feet over Murray I can often hear pilots talking to each other above an airport in suburban Nashville a hundred miles away.

After soloing I speeded up my flight training and took the check ride for my private pilot's license a few months later. Then I bought into a flying club that owned a Cessna 150 hangared in Murray. Clubs make flying much cheaper. In our case, the hourly flying rate paid by club members was a third of what it cost to rent a plane, so that after fewer than fifty flying hours the buy-in cost had been made back in savings. And a bonus for me was that in the years I belonged to the club I was usually the only one flying the plane more than a few hours a year. Even though there were eight of us in the club, it might as well have been my private airplane.

4.

Even before I got my license I rented airplanes when I was in Tucson. Flying in Arizona, where I grew up, has a special sort of satisfaction and aesthetic pleasure for me. It was the first place I flew a plane and, much earlier, the first place I flew *in* a plane. Southern Arizona has many attractions for the private pilot, not the least being the more than 300 flying days in the year. In the summer, admittedly, these may be half days, since the midday density altitude—the altitude at which your airplane's engine thinks it's operating—can reach a number close to its operating ceiling, and then it will refuse to climb.

The first time I heard on the radio an automated weather report announcing a density altitude of over 5,000 feet, I had a vivid flashback to a flight with my stepfather and my mother when I was perhaps twelve years old. One summer Saturday we flew into Prescott for a weekend in the cool alpine forests of northern Arizona. When we left to go back to Phoenix on Sunday it was a hot day. Seth topped off the tanks of the Cessna 175 and we all climbed in. Never having flown out of this airport before, Seth failed to consider the effect of the six-thousand-foot altitude and the density altitude on the takeoff roll. He pushed in the throttle all the way, the airplane rolled down the runway... and rolled and rolled some more. As the end of the runway came into view, the airplane began to inch up off the tarmac. We were perhaps five feet in the air as we cleared the runway, and then the problem was the approaching pine forest, where the trees were sixty feet tall. But once we were past the end of the runway, the plane seemed to take heart and began to climb a little faster. We cleared the trees with a dozen feet to spare.

But in the cool of the early morning in the summer and just about any time of day during the rest of the year, the Arizona skies are clear and inviting. Sometimes the spring wildflowers, orange patches of poppies extensive enough to be seen from the air, dot the slopes of mountain ranges that still have snow clinging to their peaks. I have been lucky enough, several times, to take off on a cold winter morning after a rare snowfall covered all the mountain ranges down to their skirts in the foothills, before the inevitably warming day reduced the snow to small white caps or melted it completely.

The southern third of Arizona, outward from Tucson for a hundred miles to the north and south, from California to the New Mexico border, is basin and range country. The mountains, from four thousand to ten thousand feet, are arranged in ranks that each stretch a few dozen miles across the desert; between them is the flat land of the high desert to the south, sloping from about four thousand feet at the border down northward toward Phoenix's low desert of a little over a thousand feet. The mountains often have distinctive shapes, either whole, like the massive wedge shape of the Santa Catalina Mountains just north of Tucson, or in their individual peaks. The striking form of Cochise's Head tops the Chiricahua Mountains at the New Mexico border, two heads (Dos Cabezas) look down at the wintering Sandhill Cranes on the wetlands outside Willcox, the Santa Rita Mountains have a distinctive saddle curving beneath their peaks, and so on. The result is that wherever I fly in southern Arizona, landmarks keep me located. As soon as I climb out of the pattern at Ryan Air Field, the peculiar carved shape of the massive granite block at the top of Picacho Peak becomes visible thirty-five miles to the northwest. Away to the southwest is the equally familiar Baboquivari Mountain, sacred to the Indians of this region.

The mountains do not tower once you are in the air; they subside to become interesting variations on the horizon. Always and everywhere the flying world is mostly sky. In the southwest this expanse merely amplifies an effect you see on the ground as well. My favorite quote from Willa Cather gets at the opening up of the landscape in the high desert: "Elsewhere the sky is the roof of the world, but here the earth was the floor of the sky. The landscape one longs for when one is far away, the thing all about one, the world one actually lived in, was the sky, the sky!"

Flying north from Ryan Air Field to the Picacho Mountains, you can sense the floor of the desert falling away and the temperature rising. The ascending air buffets the plane slightly, especially where the light brown beneath turns suddenly bright green on the margin of an irrigated field. Farther ahead the landmark that has been growing is the sheer edge of the Superstition Mountains, the promontory called The Flat Iron, marking the eastern edge of Phoenix's sprawl. You've got to keep up to six thousand feet to clear the highest peak here, but it's necessary to fly over the range rather than take the inviting open passage to the west. That would put you in the controlled airspace around Phoenix's Sky Harbor Airport. Bravo or B airspace it's called, the particular area around the very busiest airports in the country. You can go in there if you want to request it on the radio, but you have to be told specifically by an air traffic controller that you are clear to enter Bravo airspace, and, once inside, you'll have to follow directions about exactly what direction you fly and how high — when the air traffic controller says "Jump", you don't have to ask "How high?" because she'll tell you. Better to climb over the Superstitions outside of Bravo airspace. You will need the altitude anyway, because as soon as you're over you'll see it's nothing but mountains and valleys from here north. You have entered Arizona's central highlands, a band of many mountain ranges packed close together across the state from just north of Phoenix all the way up to the Mogollon Rim of the Grand Canyon.

5.

Flying over the Superstitions recently and beginning to cross the Salt River to wind my way up the Verde Valley toward Sedona, I had another flashback to the Prescott trip my family took so long ago in Seth's Cessna 175. To fly from Prescott down to Sky Harbor Airport in a small plane, the usual way on a clear day is to follow the Black Canyon Highway. But on this day, halfway down, clouds began to build up ahead of us, and they quickly turned into towering thunderheads. They were too high to fly over and too dangerous to fly through, so we began to dodge them. Sitting in the co-pilot's seat, I watched the few clouds ahead turn into an almost solid wall. I didn't notice that the clouds had closed in behind us. I didn't look at Seth, either; I was being entertained by our flying back and forth, since we usually flew straight and level for periods of time tedious to a teenager. When I did look at him I was shocked to see he was dripping with sweat.

Seeing those mountains on my way to Sedona and thinking about flying through their valleys to avoid lowering clouds ("scud-running" pilots call it), I suddenly realized that I had been mistaken about my early flying with Seth: he wasn't really a very good pilot. He was competent enough at the controls, coordinated, with good vision and peripheral vision, watchful for other air traffic. But being a good pilot consists mostly in thoughtful planning, and I was seeing for the first time, through these memories, that Seth hadn't been good at that crucial aspect of flying.

Other memories about Seth and flying came crowding in. One was an odd dream. One night I dreamt I was standing next to Seth, looking up at the sky as my mother, in an F-15 fighter jet she didn't know how to fly, was zooming erratically around the sky. "We've got to get her down!" I said, "She doesn't know what she's doing." "It's all right," Seth answered, "she has to learn sometime."

Another memory: shortly after he got his pilot's license, Seth took me up in the Luscombe two-seater he had learned to fly in. We climbed for a long time and then he asked me if I would like to see a stall and a spin. I nodded and he throttled back and lifted the airplane's nose sharply. It slowed, and then suddenly we were looking at the ground, which was violently turning clockwise. Seth centered the control stick, then stepped on the rudder to stop the spin. As the airspeed built up he added power and gradually brought the nose back up to level. I was exhilarated but also frightened. Seth noticed I was pale and clutching my seat because he asked if I was okay and then added, "What were you expecting?" I don't know what I thought a spin would be — perhaps something like a slow spiral.

I have to remember, though, that Seth was only thirty-five when we did our flying together. I think about those pine trees a few feet under the wheels, those clouds closing in, but most of my memories of those flights are happy ones. Pilots like to say doctors in Cessnas are the most dangerous things in the air, but I don't remember any other perilous situations. Once we were approaching Tustin and flying above clouds that had formed beneath us, then had almost completely solidified. We found a hole and dived through it. Afterwards the overcast lowered and there were no holes for days. We ended up taking a commercial flight back to Phoenix and sending someone to ferry the plane home later. That may have been the point at which Seth decided to sell his share of the plane.

We took flying trips for several years, and then Seth stopped flying. I didn't learn to fly until much later. My stepfather ultimately got me into flying after he died. I date the beginning of my flying instruction from the flight where we scattered his ashes over the Santa Ritas. Shortly before he died he gave me his car, a Thunderbird. I drove it for a while, and after he died I sold it and used the money to pay for flying lessons. It would only be stretching a little to say that, one way and another, my stepfather taught me to fly. But I wonder whether some of my flying memories didn't delay that first takeoff.

14. Selling My Library

"Have you read all these books?"

I.

This morning I packed up two books to send off to their new owners. Sometimes the combination of book title and buyer fascinates me, and that was the case with one of these books: *The Englishness of English Art* went to a buyer named Stewart-Gordon, English squared to double Scotch. In other cases I find the buyer's location oddly apt — or just odd — in combination with the book: I can imagine why a gentleman living on Bear Wallow Road in Huntly, Virginia might buy *Life in an English Country House*, but it was a man from Los Angeles who bought my anthology of New York poets, and why would a sociologist at Yale want my old copy of Ian Watt's *The Rise of the Novel*?

The other book I sold this morning was T. S. Eliot's *Complete Poems and Plays*, the hardback with the teal-green buckram cover so familiar to generations of English graduate students. Many of the books I send away have annotations and marginalia that reflect my thoughts when I first read them or slogged through them for a course or gleaned them for a critical article I was writing. But Eliot had only my bookplate inside the front cover, "Tucson, Oct 23, 1965" written on the flyleaf, and, in the table of contents, publication dates penciled in behind "Choruses from *The Rock*" and each of the *Four Quartets*.

I know that this discussion of selling my library is going to seem horrible to a certain kind of reader, like describing the dismembering of one's child. I know because I have been that sort of reader myself. But bear with me. I will reassure you by saying that I'm not selling *all* my books. But I have sold more than a thousand of them so far, and I don't intend to stop soon.

II.

I love books. But there is more than one way to love them. Anne Fadiman makes a good distinction between kinds of booklovers in "Never Do That to a Book" (*Ex Libris*, 1998). "Courtly lovers" use bookmarks, never dog-ear pages, and don't put books down with pages splayed. "Carnal lovers" not only do these things but also write in books, use them for doorstops and desk props, and generally "love their books to pieces".

Fadiman's distinction points cleverly to the uniqueness of books as objects. Books present a philosophical oddity: unlike any other objects, they exist as individual artifacts but also have a transcendent being: a book is its pages, ink, glue and binding, but it is also its content, independent of any physical copy. Ray Bradbury portrays this dual nature of books in his 1953 novel *Fahrenheit 451*. In the book's frightening picture of the future, a totalitarian government controls the populace through constant, mindless entertainment and strictly forbids any free play of ideas. Books, when found, are immediately burned (the title refers to the temperature required to ignite a book). But a group of subversives is fighting back; in Bradbury's story *people* become books, each person memorizing a favorite book because these booklovers know the physical copies will soon be confiscated and burnt by the regime.

One can love books in this transcendent sense, and I like to think of myself as a Bradburian booklover of the essence of books. Andrew Lang calls the other sort of love of books *bibliomania*, which he defines as loving them "for their own sake, for their paper, print, binding, and for their associations, as distinct from the love of literature" (*Books and Bookmen*, 1887). To love books in this way is to be obsessed with their most worldly aspect, and I know because for many years I had the mania, or to put it in coarser words, I was a bookrat.

For most of my adult life I collected books, as bookrats do. I have traded for books, bought them new in bookstores, ordered them from publishers and book-shops. In the old days—the sixties and seventies—a new English book from Blackwell's in Oxford, shipped surface mail to the States (which took about six weeks), was still a third cheaper than if bought here. Those new expatriate books were somehow more exotic, but every new book has its aura, its smell of scarcely-dry ink, its virgin feel of new binding that's never been stretched. Not that old books lack their sensual charms, being worn to the hand, with pages that turn easily and lie comfortably. The erotics of book-owning is a subject yet to be explored.

I have no idea how much I've spent on books over the forty-five years or so I've been buying them. I imagine the number is at least fifty thousand dollars, and it could be twice that. But a bookrat is liable to think of money spent for books as no more to be tallied and examined than the cost of a needed transfusion for an injured family member.

It's painful for a bookrat to let any book go. This retentiveness is as familiar to another bookrat as it is puzzling to one who is not. The bookrat has an answer when the booknaif asks, "Have you read all these books?" In fact, the bookrat has a whole series of retorts. "Yes [a small lie], and I may want to read them again." "Not all books are to be read [a rationalization]; some are just for reference or judicious sampling." "No, but I may want to read any one of them any day [the whole truth]." "I just feel more secure knowing they're there to go to [and nothing but the truth]." Anatole France's answer was "Not one-tenth of them. I don't suppose you use your Sèvres china every day?"

In his *Philobiblon*, written sometime in the first half of the fourteenth century, Richard de Bury uses Aristotelian logic to demonstrate that since wisdom is contained in books and wisdom ought to be prized above all things, we should all be lovers of books. For most bibliophiles, reason isn't in it: booklove is a passion. And yet it's hardly ever really promiscuous. I don't have stacks of science-fiction books as my daughter-in-law does, or the espionage thrillers that my younger son has shelves of, or out-of-print bird books such as my wife collects. What I have mirrors my taste, and this reflection may be another reason it becomes harder to get rid of books the more we have: a large collection of books is a portrait of its owner, and it is hard to part with a flattering image of ourselves. How could it not be flattering? It's made of *books*.

III.

Milton's *Areopagitica*, written three hundred years before Bradbury's *Fahrenheit 451*, also presents us with a metaphoric identity of books and people. But where Bradbury tells a story in which people can become books, Milton turns the metaphor around:

> For books are not absolutely dead things, but do contain
> a potency of life in them to be as active as that soul was
> whose progeny they are; nay, they do preserve as in a vial
> the purest efficacy and extraction of that living intellect
> that bred them. I know they are as lively, and as vigorously
> productive, as those fabulous dragon's teeth; and being
> sown up and down, may chance to spring up armed men.
> And yet, on the other hand, unless wariness be used, as

good almost kill a man as kill a good book. Who kills a
man kills a reasonable creature, God's image; but he who
destroys a good book, kills reason itself, kills the image of
God, as it were in the eye. Many a man lives a burden to
the earth; but a good book is the precious life-blood of a
master spirit, embalmed and treasured up on purpose to a
life beyond life. 'Tis true, no age can restore a life, whereof
perhaps there is no great loss; and revolutions of ages do
not oft recover the loss of a rejected truth, for the want of
which whole nations fare the worse.

Killing a man is bad enough, but killing a book
might just be worse. The liveliness of a book, the "po-
tency of life" in it, consists not merely in its presence
in one mind, but in its dispersion, its *reproduction*, "as
vigorously productive, as those fabulous dragon's
teeth" which "being sown up and down, may chance
to spring up armed men." Books are vigorously productive to the extent that they circulate.

Surely the most bizarre use of the living book trope
is the conversation Washington Irving's Geoffrey Crayon has with a "little tome" in the library of Westmin-
ster Abbey (*The Sketch Book*). "How much better you
are off," Crayon says, "by being stored away in this
ancient library?" But the book will have none of it. "I
was written for all the world, not for the bookworms of
an abbey," says the book. "I was intended to circulate
from hand to hand."

IV.

My thinking has gradually changed from that of
the bookrat's retentiveness to a different attitude about
books. I still want some books around me, but where
the promiscuous bibliophile wants books of any sort,
at one point I realized that a lot of the books on my
shelves were trash and should go. They were freeload-
ers taking advantage of my hospitality. There were art

books that didn't describe their subjects or provide good pictures of them or present interesting opinions about them. There was bad poetry; there were worse novels. I began to think about culling.

Being forced to get rid of some books probably helped change my bookrat's feeling that all books are precious. Chronic book-retention syndrome has to give way to life-changes and moving days. When I married, my wife Katharine and I were both in English graduate programs, so there was much duplication among our books. But even getting rid of the duplicates was hard. Anne Fadiman has another wonderful essay entitled "Marrying Libraries", a title that will immediately evoke a flash of recognition from any pairs of bookrats who have ever had to decide whose annotated copy of *Don Quixote* or *Giles Goat Boy* was going to be pitched. Marriage itself pales, Fadiman writes, beside "the more profound intimacy of library consolidation." But we did cull a lot of duplicates, and at the same time we managed to get rid of a few additional books on each side.

Twice we moved our books, along with the whole household: once from graduate school to New Orleans, where I taught for a few years, and then again to the west Kentucky town where we've been for thirty years. Each time we tried to cull books, and each time we thought we'd done a good job. But boxing the rest of the books, carrying the boxes, loading them in a van, unloading them, unpacking the boxes—all of this reminds the aching backowner/bookowner that there are still too many books.

My wife's library and mine have developed separate specialties and have redivided over the years; her guidebooks and works on natural history occupy their own rooms, while my books on painting and astronomy accumulate in my study. We both like mysteries, but rarely read the same authors, and my collection of mystery books gradually turned into a small research library. I also had a professional library of books on language and literature in my college office.

Then I had to switch offices. The book space in the new office was less than half that in the old; I had to do some drastic culling. I let my colleagues pick over the culls and discovered that giving away books to them was much easier than I had anticipated. So when my retirement neared I found it easier to face the prospect of another huge culling, since my home bookshelves could hardly accommodate a dozen more books, let alone the hundreds on my office shelves. This time I thought I'd try to sell some books.

V.

I began in a big way: I would try to sell my 500 volumes of mystery and detective fiction all at once. These books were the primary source materials for the book I wrote on the appeal of mystery fiction, *Murder Most Fair* (Fairleigh Dickinson, 2000). Looking at eBay, I made the disheartening discovery that groups of books seemed to be practically worthless. I tried auctioning them on eBay nonetheless, but no one got even close to bidding my reserve price, which I thought ridiculously low.

I decided on another strategy: I had some email lists from those who had participated in mystery fiction panels with me, members of popular culture groups, and a nineteenth-century literature study group. I sent an email to everyone. I made it as attractive as I could, and I set the price ridiculously low. Most of the books were paperback reading copies, but about a hundred were hardbound, including some first editions and a nine-volume Oxford Sherlock Holmes in mint condition. The collection was very comprehensive, spanning almost all the titles in various lists of best mysteries such as H. R. F. Keating's *Crime and Mystery: The 100 Best Books* and the Howard Haycraft-Ellery Queen Definitive Library. Included was all the longer fiction and most of the collected shorter fiction of Dashiell Hammett and Raymond Chandler, as well as most of the works by Manning Coles, Dorothy Sayers, G.K Chesterton, and Sara Paretsky. Significant English and American writers from the nineteenth and twentieth centuries were there, as well as some writers from Australia, South America (Taibo, Vargas Llosa), the Far East (Natsuki), and Europe (Simenon, Arjouni, Sjöwall and Wahlöö, and others). And because I was not sure the number of books would convey it, I was careful to specify that the collection amounted to twenty-eight shelf-feet of space.

I got one reply. Amazingly, though, that one respondent went ahead to buy the books. She was a graduate instructor at the University of Virginia, having just finished her Ph.D. there. She described herself as "a recent Ph.D., now budding mystery novelist", and she was excited about acquiring the books. She not only had an interest in mysteries; she was beginning to write mysteries herself. Her interest and her situation made it easier for me to part with the books. While I addressed cartons to her in North Garden, I was thinking of her reading my Poe stories "while residing near Charlottesville, Virginia", like the narrator of Poe's "A Tale of the Ragged Mountains." I mailed off the books in a dozen boxes. It was September, the beginning of my last year of teaching.

VI.

I had been lucky to sell the mystery library so easily. I knew that the remaining books would have to be sold

individually, and I discovered there were several ways
to do this using the internet, which is far and away the
most efficient way to sell books these days. One can
become a dealer and lease advertising space on the
net; these services assess a monthly charge according
to the number of books one has listed. I didn't know
at the time how many books I wanted to sell. What I
needed was an internet service that would list as few
or as many books as I wished, and one that did not
charge until a book sold. That describes Amazon.com's
mail-order book business called Amazon.com Market-
place.

Amazon runs the perfect American middleman
scheme: they don't handle books, they don't house
books, and they don't mail books. They just collect the
money, take a hefty fifteen percent cut of it, take *another*
cut out of the amount the buyer pays for mailing, and
put the rest in your account. It's all done electronical-
ly. They list your books for two months, and if they
have not sold by that time, they notify you and you
may relist if you wish. It costs you nothing to list the
books, and Amazon's commission is only subtracted
if the books sell. You are responsible for mailing the
books within two days of their sale.

They make it tremendously easy for you. All you
do is find on the Amazon.com website a page headed
"Sell an Item". There you can enter either a title or the
book's ISBN number. Amazon tells you whether anyone else is selling the book and what the lowest price is;
if you wish you can browse all the listings. You decide
whether to try to get an average price for the book or
to undercut everyone else. You are responsible for an
accurate description of the book. The overall categories
are new, used (with a number of subcategories), and
collectible, also with subcategories. If a book is a first
edition, bears an autograph, or is collectible for some

other reason, you may list it for whatever price you wish; otherwise you may not charge more than Amazon's retail price for the book.

VII.

I started to sell books on Amazon.com a month or so after I sold the mystery books. In the first month I sold only four books, but I didn't list many. Once my listings increased I began to sell books more often. By February I was selling a dozen books a month. By the summer I had 250 titles listed, and I settled into a routine of selling about twenty books a month.

I have sold books to buyers in every state in the Union, and some of my books have gone to Germany, to France and Italy and Spain and Norway and the Netherlands and England and Ireland and Brazil and Japan and South Korea. I've sold books to people in Boring, OR, Loveland, CO, Hernando, MS, Nevada, MO, and Grottoes, VA. I have repeat customers; one man looked at all the books I had for sale (one can view them on the Amazon website) and bought several books of Jorie Graham's poems. I infer from some of the addresses in Austin, Ithaca, Bloomington, Ann Arbor, Columbus and other college towns that college students are buying some of my books. Some college libraries also have bought them: last week I sent off a copy of Weldon Kees's novel *Fall Quarter* to the Sarah Lawrence College Library, and some time ago the Beeghly Library at Ohio Wesleyan bought a copy of Christopher Wood's *Victorian Panorama*. The National D-Day Museum in New Orleans, which the historian Stephen Ambrose had a hand in starting, bought the copy of *Crazy Horse and Custer* that Steve had autographed for me when we taught together at the University of New Orleans years ago.

Some of my books were priced low enough that they were bought by dealers, or perhaps they bought my books to fill orders for special customers, and the price didn't matter. But most, I believe, have been bought by people who are neither students nor librarians nor dealers, those whom Virginia Woolf and Samuel Johnson called common readers.

VIII.

Francis Bacon wrote that "some books are to be tasted, others to be swallowed, and some few to be chewed and digested." My friend Bob Bourdette would add that some are to be thrown violently across the room (I watched him react that way to the cant of *Zen and the Art of Motorcycle Maintenance*, which he read during its fifteen minutes of fame). I would add that some books are to be sold — after we have had them for a while, of course, and have read them several times over and know we won't be reading them again. Or perhaps not have read them and know we never will.

Books are artifacts, sometimes works of art, but they are also commodities. Given their dual nature, it is not surprising that Walter Benjamin, the author of the influential essay "The Work of Art in the Age of Mechanical Reproduction", should have things to say about the book's duality. And he does, in a very personal essay written in 1931 titled "Unpacking My Library". Benjamin begins as his own books, which had been crated up for several years in storage, are now half-unpacked in disarray around him, but will soon be shelved "in the mild boredom of order." Thus he starts with a contrast between disorder and order: libraries, or at least their catalogues and shelves, are orderly; books, full of revolutionary ideas that defy classification, are not. This tension between order and disorder characterizes the life of the collector as well, since he tries to capture books and to hold onto them as objects when their real life is not material. It is "a very mysterious relationship" between the collector and his books, and Benjamin's populist politics, he knows, are very much at odds with his love of collecting books.

Benjamin half mocks the notion of the collector sitting on a huge and valuable book-hoard (there is irony here, since his own "collection" was no more than 2,000 books), perhaps having inherited it ("inheritance is the soundest way of acquiring books"), and passing it on to heirs. This picture is an anachronism, he says. Moreover, he thinks the collecting urge is a kind of boyhood impulse. He knows that he is writing about the valuing of books for reasons — their association with places where he bought them, for instance, or their comparative value, especially when he acquired them for less than their market price — reasons that have nothing to do with the real value of books as instruments of democracy that decentralize knowledge, authority, and power. This real value of books, "their functional, utilitarian value — that is, their usefulness", depending on circulation and wide availability, is at odds with the collector's habit of hoarding and his valuing of rarity.

IX.

"If a book can be bought, it can also be sold. A book that I once wanted I may want no longer." Since I can say these things with conviction, I must have ceased being a bookrat. I have not stopped appreciating copies of books as aesthetic objects or even as repositories of memory. But I can imagine a book moving from my possession and becoming the permanent possession of someone else. Well, I don't have to imagine it; it happens every other day.

153

A man in upstate New York has bought my Hughes
edition of Milton. I look through it before packing it
up and I find oddly few notations. I read almost every
word in this book during a graduate course in Milton.
There are a few underlinings in *Paradise Lost*, though
none in *Areopagitica*. I ask myself what I feel at letting
this book go. I recall that it wasn't a course I was pas-
sionate about, and yet I did good work in it, writing a
paper on a manuscript question in *Comus* and another discussing the occasional rhyme in the blank-verse
closet drama *Samson Agonistes*. The paper on *Samson*
was published by *Milton Quarterly* and was one of my
very first publications. The marginal notes in *Samson*
are some of the very few annotations in the book. I haven't looked into this book in perhaps twenty years.
What I feel is that this book is ready to go to its new
owner.

A. R. Ammons' *Collected Poems* is going to a woman
in the Department of Ecology and Evolutionary Biology at Cornell. I imagine her reading "Corson's Inlet",
with Ammons comparing the swerves and eddies of
his lines to the "inlet's cutting edge". Is she interested
in the descriptions of bayberry and reeds, the "black
shoals of mussels", or perhaps the interplay of form
and chaos, order and entropy in one day's observations
of the inlet? Has she also ordered Robinson Jeffers? I
wonder. A woman at the McDonald Observatory in
Texas has bought Copper Canyon Press' anthology of
poetry that marked their twenty-fifth year of publishing; I imagine her sitting at the control panel, reading
poems on cold, clear winter nights while the big scope
takes digital images of galaxy clusters a billion light
years away. Sometimes I wonder whether I should tell
her about the pharmacist in Biloxi who likes Wallace
Stevens, or the man in New Jersey who bought those
three books of Jorie Graham. But I won't, and I won't

tell the woman at the law firm on Times Square who bought J. L. Austin's *How to Do Things with Words* that she may find Austin's philosophic theory of speech acts doesn't easily translate into effective closing arguments.

I know that these attempts to read the future of my books mean I'm trying to hang on. Some part of me still believes I should only acquire and never get rid of books. But most of me knows otherwise. My library when I began selling books was only a little bigger than Walter Benjamin's. Now it is considerably smaller. I now spend more time in the city library than I used to, and almost as much time in college libraries as I did before I retired. I am learning to love libraries again in the way I did before I was able to afford any book I wanted. I have open bookshelves. I have freedom. And I have all this money. What books shall I buy?

15. Watching Birds and Stars

Last night I heard three different owls while I was watching stars. All were invisible in the moonless dark, though easily identified by their calls. *"Who cooks for you? Who cooks for you all?"* barked the Barred Owl, while the Great Horned Owl uttered a deep *"who who hooooo who who?"* Next, the Eastern Screech-Owl gave its startling combination of a whistle and a scream.

At first blush you might think watching birds and watching stars are as different as night and day. Certainly the differences seem more profound than any likenesses. The brief lives of birds one catches on the fly, with their vivid mating colors, their frantic activities, and their sometimes miniscule size even magnified in binoculars — these have little in common with the enormous extent or the muted colors, mostly whites and grays, of deep-sky objects such as galaxies and nebulae. Even the stars, though colored icy blues, oranges, and reds, are austerely tinted, never gaudy. Magnitude typifies stargazing, with the staggering distances and clusters of billions upon billions of celestial bodies, and the adamantine fires of unimaginable heat, compression and gravity beyond conception, beyond belief, not to mention the ages of these wonders, starkly contrasting with the short lives of birds.

But likenesses there are, not in the objects of the watching, but in the watching itself. For starters, the equipment for basic watching is the same. In day-to-day observation, the naked eye suffices to see birds and celestial objects as well. A little magnification is good, and I use the same 10 x 42 binoculars for most of my birdwatching and much of my skywatching.

But the most important observing instrument for either activity is the mind; birding and skywatching are linked by the need for a proper state of mind. One has to be receptive to see, and receptivity means patience above all. "Go with me just a little farther up the trail," my wife Katharine said to me in Arizona's Madera Canyon one May morning. It was hot, the trail was rocky and steep, we'd seen no more than three birds in our quarter-mile hike from the trailhead, and I was thirsty and a little dizzy from having had no breakfast and being at a higher altitude. I acquiesced, grudgingly, and when we had climbed another dozen steps, an Elegant Trogon, flashing Christmas reds and greens, flew across the trail and perched, posing, in an open spot in a conifer. My wife's near Zen-like patience when she is looking at birds comes from her clear awareness that there is nothing else she would rather be doing. When she glimpses a bird she can't quite identify because it moves away, she waits. Almost always her patience brings it back for another look.

You would think the ancient galaxies and the slow-moving planets could not withhold their beauties because of the impatience of an insignificant watcher light years away. But in fact the impatient skywatcher sees little of what the heavens offer. One cannot even begin to see dim objects until the eyes are dark-adapted, so the first twenty to thirty minutes under the stars have to be indirect gestures toward seeing: gathering charts and eyepieces and aligning a telescope so that it can track the stars in their slow, high arcs, all the while using only a dim red light that lets the eyes adjust to darkness. The actual observing can't be hurried, either. When watching Jupiter or Mars, I see at first view only a disappointing blur, lacking any detail. To truly see one of these planets requires spending many minutes looking through the eyepiece just to be rewarded with a few seconds of a clear view of a band of clouds or a polar ice cap.

Birding and stargazing require an element of concentration that goes beyond mere patience. One has to be prepared to see what is to be seen, not necessarily what one has planned to see. A certain calm is required. It's as if, when one is looking for birds, agitation of mind warns the birds away.

When at a telescope, star-hopping — using the charted arrangement of stars to move from a known bright star to an object one is seeking — becomes practically impossible when one is impatient. The technique requires finding a known, reasonably bright star using the naked eye, locating it in the eyepiece, and then comparing the telescope view of the dimmer stars around it with a chart of the sky. Then one slowly makes one's way through the pattern of stars toward the star cluster, galaxy, or nebula one seeks. The trouble comes when one gets off course in a complicated starhop — usually requiring a return to the starting point. There are times when I come back to look for a particular object after I've been frustrated by my failure to find it, sometimes for several nights or several watching sessions, and all of a sudden I get where I want to go and find what I'm seeking with no effort whatsoever. I aim the finder-scope (a small, low-power telescope aimed at the same place as the higher-powered scope and used to point it), look in the eyepiece, and there's my Smoke Ring Nebula or Whirlpool Galaxy in the field. The process seems more intuitive on the nights of my success. There are "go to" telescopes that do the work of finding for you; just plug in the sky coordinates or in some cases just enter the name of the object and the telescope whirrs, pivots, and takes you to the object. It's much like birdwatching in an aviary.

But the real seeking, either birds in the field or stars in the sky, is a patient and receptive activity that shades off into the meditative or even spiritual: birding and stargazing are humbling experiences that take us beyond human space and time and enable us to get closer to, if not an understanding, at least an awed regard for space and time as they are experienced by other creatures on this planet and by the non-eternal stars.

We have to slow down to see either birds or stars. The birdwatcher's productive rhythm is *slow step, slow step, stop, scan*. The usual rhythm of walking, which is faster than your heartbeat, must be slowed to a rate about that of your respiration, punctuated by full stops: *slow step, slow step, stop, scan*. Sometimes the rhythm is no rhythm at all, but stillness. *Sit still, wait.* Wait just beyond the point of certainty that waiting is futile; the bird I just saw out of the corner of my eye will *never* come back. Then wait a little longer. Sometimes one must be immobile for many minutes. This is the only speed at which one can see birds, but it is not the speed at which they live. Because they live faster, briefer lives, we have to slow down to see them.

If birdwatchers look as if they were moving in slow motion, stargazers at their scopes appear not to move at all for minutes at a time, as if frozen at their eyepieces — on winter nights an impression too close to truth for comfort.

Both these activities sensitize us to the daily spin of the earth into and out of darkness and its larger, yearly wheel around the sun. Autumn means the return of white-throated sparrows and the pleiades. In winter Orion is high in the sky after dark and, during the day, grosbeaks, cardinals, and sparrows cluster at the feeder and peck at the snowy ground. In the spring, the warblers arrive and the early evening sky presents its array of galaxies in Virgo, Coma Berenices, and Leo.

Summer means whip-poor-wills singing during nights dominated by the Lyre, the Swan, and the Eagle. All of these are real days and nights to mark one's life.

***

Last June Katharine and I went on an Audubon Society field trip to the Coyote and Quinlan Mountains, southwest of Tucson. It was hot that day, perhaps 104. The trip ended on Kitt Peak, at 6875 feet the highest point of the Quinlan Mountain range. On top of Kitt Peak is a compound where a consortium of universities operates a number of telescopes. We stopped in the parking lot and put on the gear we'd been wearing earlier: the binoculars, the fanny packs with the necessary water bottles, and the bird books. We walked through what the observatory consortium calls "The World's Largest Array of Optical Telescopes", looking in piñon and mesquite, scrub oak and manzanita bush for birds. Mexican Jays were there on the peak, with their smooth dusky blue, uncrested heads. We saw the bewick's wren, not so much to look at, but with a remarkable song that sounds like an old rotary phone dialing nine. We saw Chihuahua Ravens and White-throated Swifts flying overhead, and we saw the little Bridled Titmice of the higher elevation in the West, as well as Brown-crested Flycatchers and Acorn Woodpeckers with their black, red, and white clown faces.

We were there for birdwatching, but we agreed that there was one telescope we ought to look at, the Mc-Math-Pierce Solar Telescope. It is the world's largest solar instrument and doesn't look like a telescope at all. On the very top of the mountain and visible from fifty miles away sits an odd-shaped building with a tower nearly one-hundred feet tall and a diagonal shaft slanting down into the ground. The arrangement looks a bit like the huge diagonal pulley belts that bring copper ore up to the top of a building that houses a rock-crusher — such structures can be seen around Arizona's open pit copper mines. In the solar telescope, the diagonal shaft is the telescope's light tube, and the shaft is continued by a tunnel cut into the earth some 280 feet. Visitors can go inside the solar telescope. To reach the ground-level door, we walked under the arch formed by the diagonal shaft and the upright tower. Suddenly the southern Arizona heat was tempered with a cold breeze. Because hot air moves and distorts light passing through it, the builders cooled the metal skin of the diagonal with circulating water in pipes beneath its metal skin, and wind flowing over the skin is cooled in its turn.

We entered the telescope just at the point where the slanted structure is continued underground with an angled tunnel cut into the top of the mountain. Above us, at the top of the shaft, we could see the three mirrors that turn with the sun's movement and keep its rays shining down the length of the diagonal shaft and the tunnel

Minutes later, we emerged from the solar telescope shaft into the full force of the high desert sun, walking through the cool arch and down the road past the dormitories where the observers who used the other telescopes, the nighttime ones, slept during the day. In the trees near the dormitories one dusky-capped flycatcher was singing, and we walked over for a closer look. Standing in the shade of the oak, we watched the flycatcher dart from his perch to catch an insect, return and sing again. Looking down, I was startled to see the ground covered with tiny butterflies, the varieties called Spring Azure and Marine Blue, both of them seeming to reflect the intense hue of the sky. They clustered at a tiny damp spot on the ground. Male butterflies congregate on moist soil or mud to extract mineral salts, compounds of metals that had been made in the heart of distant stars, older than our sun, whose young thermal furnace fuses only the lightest element, hydrogen, into the next lightest, helium. The butterflies, nearly motionless, sucked the moist ground in the shadow of the solar telescope, but around them tiny images of the sun moved with the gaps in the leaves as the tree branches swayed in the light breeze.

16. A Round with Friends

The only way of really finding out a man's true character is to play golf with him.

　– P. G. Wodehouse, "Ordeal by Golf" (1922)

1.

One day, tired of the cheap, plastic feel of the Top-Flite golf balls I had been playing, I told my long-time golfing partner, Richard Steiger, that I had decided to go back to playing with a wound ball. "Was you using a sqwuawe one?" Richard came back, without a pause and in his best Elmer Fudd accent. Richard's quickness on the uptake is legendary among his friends, and he is often at his funny, cheery best on the golf course. He has the most inventive excuses for missed shots that I have ever heard. One summer day I watched him attempt to hit a shot from the rough. Bees buzzed in the blooming clover surrounding his ball. He topped the shot badly. "The bees rushed me," he explained as he got back in the cart.

　We were on the first tee of Miller Golf Course, built in 1983 by the college where Richard and I both taught English until our retirement. Richard had provided names for each hole. The sixteenth, shortest of all the par three's, was "Happy Fun Hole", the fifth, with its impossible to hit, saddle-shaped narrow fairway was "Hole o' Death", and seventeen, with its elevated tee, was "Golan Heights".

　It was on account of Richard that I took up golf again after a long hiatus. Richard's regular weekend game had been tennis, on the courts where he had a reputation for salty language and opponent-baiting. But after his heart attack, when he had recovered enough to get some exercise, he found tennis was going to be too strenuous. He remembered a college P. E. course where he'd enjoyed hitting golf balls (at CCNY they never actually took the beginners out on a golf course). When he discovered I used to play golf, he suggested we go out together.

Richard enjoyed himself thoroughly. Never much of a nature lover, he found this tamed and contoured variety perfectly to his taste. Although he would utter an occasional expletive if he missed the same shot several times in a row, he stayed cheerful, pleasant, and, best of all, witty. When I told one of his former tennis buddies about the new Richard, he was amazed; then, emulating his subject, he quipped: "Ah, Richard. Coarse on the courts and courtly on the course."

2.

In July of 1999, my stepfather and I were on the second tee of Haven Golf Course in Green Valley, Arizona. As I turned away from the ball in my backswing, I felt the first raking pain in my hip that would eventually force me into my second back operation, two months later. The effect of the pain was that I limited my turn as I swung back to the ball, "blocking" the shot and causing the ball to fly to the right. It bounced on the summer-hard fairway and scooted under some trees in the right rough.

When I got to the ball, I found it lying on pine straw under some trees we call Tamarisks in Arizona. These trees are more widely known as Salt Cedars, and as I stood under them, inhaling their dry and slightly camphoric odor, I realized that this smell meant golf itself to me. It was the smell of the fairway margins at the San Marcos Golf Course, where I had played my first round.

I was nineteen years old when I took up golf, and I did it to spend more time with my stepfather — or more accurately, to get him to spend more time with me. Seth was a small-town family doctor at a time when being "on call" meant not only taking the phone call but making the house call. He didn't have much time for leisure, and he headed for the golf course when he did. I spent some time at the practice range, hit whiffle balls in the back yard with a nine-iron, and studied Tommy Aaron's book, *How to Play Your Best Golf All the Time* — still, I believe, one of the best books for a beginner's instruction. Finally Seth decided I was ready to play a real course, and we went out one day to the first tee of the San Marcos Hotel's course in the little town of Chandler, where we lived. I promptly hit my drive off the toe of the club into the trees bordering the first fairway to the right. I spent a lot of time under the trees that day, hitting from the dry needles dropped from the Tamarisks lining most of the fairways. I came to associate that smell of dead Tamarisk needles with the game itself.

The San Marcos course was a classic Arizona layout: flat, with Bermuda grass fairways and greens. Sand bunkers and a few ponds had been added for interest, but the main challenge was negotiating the straight or doglegged fairways defined by the smoky gray-green Salt Cedars, with their brown mats of fallen needles underneath. I have since learned that the San Marcos Hotel was the first golf resort in Arizona, and its course was the first in the state to have grass when other courses were rolled sand.

Haven Golf Course in Green Valley, a hundred fifty miles south of Chandler's San Marcos course, was very like it in its flatness, its added water and sand, and especially in the Salt Cedars that edged the fairways. As I chipped out from the ones to the right of Number Two fairway, I thought about how my plan for golf to bring Seth and me together had taken a long time to work, but had finally succeeded.

Tired of the drudgery of always being on call in his Chandler practice, Seth had started a residency in physical medicine that took him and my mother to California. I was in college, then graduate school, then raising a family. Seth and my mother divorced; both remarried. He ran physical medicine and rehabilitation departments for the Veterans Administration, first in Long Beach, then Tampa, and then Salisbury, North Carolina. I taught in New Orleans, and then in western Kentucky. I had stopped playing golf regularly. We would visit with the kids for a day, but we rarely played golf. Then he retired and moved back to Arizona, and my wife and I began spending more time in Tucson, where we had met and where her parents owned a house. Eventually, golf really did form the sort of bond I'd hoped it would between us: it gave us an excuse to be together often, and we found we enjoyed each other's company.

3.

In 1973 I was playing Pontchartrain Golf Course, since renamed after its builder, Joe Bartholomew, who also built two golf courses in New Orleans's City Park, though he was never allowed to play on them. He did play on this course, though, which sits in the shadow of the traditionally black college called Southern University. I had played two holes and I was on the tee of the third when I suddenly realized that golf wasn't fun anymore. I had been playing for only half an hour but

my shirt was completely sweat-soaked. I was hot and
sticky and for all of those thirty minutes, I realized, I
hadn't been thinking about playing golf but about my
new job teaching at Louisiana State University in New
Orleans (now the University of New Orleans). Suddenly golf seemed pointless. Why play it if it didn't
provide escape from thinking about work? I walked off
the golf course and didn't play again for fifteen years,
aside from our twice-yearly trips to Nebraska, where I
would play one or two rounds with my father-in-law.

4.

Most of Beatrice Country Club's many trees had
been planted by Herb. The tee shots on the holes taking
off from the clubhouse were especially daunting, and
the worst was the fourth. After I hit trees in two successive attempts to get back on the fairway there one
day, I said, "You know, Herb, I'm going to tell you how
I feel about these trees of yours. Do you know what the
raccoon said when he was screwing the skunk?"

Herb looked at me sideways. He had told family
members that I was "awful quiet" on the golf course,
and he didn't quite know what to make of this beginning. "No, what'd he say, Mike?"

"I've enjoyed about as much of this as I can stand,"
I said, and Herb laughed politely.

In those days, Herb was regularly shooting below
par on the old course. Then the club members decided to enlarge the course grounds to include housing,
and they remodeled the golf course, keeping about
half the original holes but creating new ones out on
the farmland they had acquired for the expansion. The
new holes had few trees but were bordered with native grasses allowed to grow tall, and the course was
lengthened about five hundred yards. The renovation

coincided with a decline in Herb's health. The course was closed for almost two years, and when it reopened, Herb not only hated the changes, but was unable to play nearly as well he used to.

5.

The only round I have ever played when a caddie carried my bag was with my old friend Dusty Doster at Congressional, where his father-in-law was a member. We were visiting his in-laws' Washington area home in the summer of 1966, and Admiral Bell suggested we play Congressional. Congressional! The course had just hosted the 1964 U. S. Open. Walking up to the first hole, Dusty and I realized that we would have to tee off watched by the whole rank of caddies lined up above and behind the tee. I don't even recall my first shot, but I do remember that I began to overcome my nervousness on the fifth hole, a par five where I knocked my second onto the green with a fairway wood, and my caddie nodded in approval.

6.

My older son Matt and I were playing the Navy's Cheatham Annex Golf Course in Williamsburg, Virginia. It was October, 1999, and I was still in a back brace from the disk repair surgery I'd had in August. The sixth hole is a short par four, less than three hundred yards, and I managed to par it, the only one of the nine-hole layout I parred that day. Matt parred it as well, but then he was the club champion, and would remain so for two more years, until he finished his Ph.D. at William and Mary, moved to North Carolina, and began teaching at Duke.

7.

Matt and I were playing Silverbell Golf Course, one of Tucson's city courses, where we had been paired with a jolly Mexican named Frank and a dour Presbyterian named Lee. On the seventh hole, I hit a pretty good drive. Matt teed up next and hit a better one. Frank said, "Nice ball. In fact, nice pair of balls", and Matt said without a pause, "Thanks, I inherited them." I cracked up, and after about four beats, so did Frank. Lee just looked puzzled.

Being paired with other golfers is a little like having relatives; you don't get to choose them and you're pretty much stuck with them for the duration. One especially annoying fellow Matt and I played with kept saying "And the crowd goes wild!" every time anyone hit even a fair shot.

My younger son Dan gets a lot of curious pairings in Nashville, where he works as a guitarist and songwriter. He amuses us describing the woman with the twenty-second swing and the man who, when he hits a stray tee shot, has another ball out of his pocket and teed up before the first one hits the ground. Dan marvels at how golf often seems to make the world smaller. One day he was paired with a middle-aged man, obviously a native Tennessean, who introduced himself as "Bob, a truck driver." A third member of their group approached and gave his name as Lee Duc Tranh. The truck driver, without missing a beat, said, "Nice to meetcha, Lee Duc; I'm Bob and this here's Dan."

8.

We were on the par-three eighth at Skyline Country Club in Tucson, where Herb was a member and played regularly in the winter. It was March, 1970, and my wife Katharine was five months pregnant with Matt;

the only way she could swing the club was in a flat plane around her stomach, so she was hooking the ball around the course. On this 125-yard hole she took a three-wood and hit the ball low toward the elevated green. The ball hooked into the hillside, bounced, and rolled up out of sight toward the center of the green. When we reached the green her ball was nowhere in sight.

"Maybe it went in the cup," Herb said with the optimism of a man who's made *eleven* holes-in-one. He walked toward the pin.

"Yeah, sure," Katharine said skeptically.

"Well, lookee here," Herb said, grasping the flag and pulling it carefully from the hole. Katharine's ball rattled into the bottom of the cup!

If you believed in prenatal influence, you would have to consider Matt's ace-in-the-womb a harbinger of his golfing career. Katharine gave up golf when the boys were small, and Matt has not had an ace so far by himself, but his game is near scratch when he's in practice, and it's only a matter of time before he has one.

Herb was playing with his regular foursomes in Nebraska and Tucson when he hit all his holes-in-one; no one else in the family witnessed any of them. Eleven seems an amazing number, but then he hit the ball straight, he knew how to pick clubs, and he played four or five times a week for sixty years or more.

Richard Steiger was playing with me when I got my first hole-in-one, on the last hole of a par-three course in Murray. Dan was there for my second, on the sixteenth hole at Miller, the one Richard calls "Happy Fun Hole".

My third hole-in-one came when I was playing alone. It was an October afternoon and the light was already failing. I planned to play nine holes, but after taking a triple-bogey on the first hole I almost truncated my round by playing back up number eight, which parallels number one and leads back toward the club-house. At Miller I sometimes played just three holes—one, eight, and nine—a hilly two-thirds of a mile just calculated to get the cobwebs of the workday out of my mind and the kinks out of my joints and muscles. But the second hole is a par three, and I decided to give it a try; I could always cut across to eight and nine if the poor play continued.

The pin was back right, and I hit a four-iron into the slightly opposing wind. The ball took off fairly low toward the center of the green and started to move to the right just before it hit. After landing just on the front of the green, the ball ran back toward the pin. I lost sight of it for a second, but when I looked again, there was no ball visible on the green. It might have run off the back, of course, but *I knew it had gone into the hole*.

My heart sank as I realized that a hole-in-one while playing alone is a hollow triumph, indeed. There are no bragging rights, and worse, smug reactions ("Oh, sure you did") from those who've heard tall tales of golfing exploits before. As it happened, I recognized one of the young men on the next tee and called to him. He hadn't actually seen the ball go in, but he walked back to the green as I came up and we both saw the ball was in the hole. He was willing to attest to it on my card, but it hardly seemed to matter to me; it was as if it hadn't really happened.

I have played a lot of golf by myself. Probably anyone who has at one time had a passion for the game has had to be often on the course alone. I believe there are things about low-handicap play that one can only learn without other voices to distract from the one inside your head saying, "Look at this, remember that, pay attention." But although golf can be mastered alone, and to some extent enjoyed alone, it isn't ever as enjoyable as when it is played with friends and loved ones.

I realized, years later, that the moment of profound dissatisfaction with golf I'd felt on that New Orleans course in 1973 was not merely about the game's losing its power to distract me. I had no one to play golf with there; naturally I was dwelling on my own concerns. Golf is finally a game to be played with one, two, or three people you want to spend time with.

9.

"I feel calm and strong," Dan said before we played El Rio the day in 1999 he first broke eighty there. Three days before, Dan had had his first score in the seventies, a seventy-nine at Silverbell that included a one-over-par back nine.

Dan, like Matt, has had trouble with anger on the golf course. They get it from me, of course, but each is unique in his approach to the game. Dan's intense desire to score well comes from perfectionism rather than competitiveness. Matt and I want to get the ball into the hole in fewer strokes than our opponents or ourselves in previous tries, and we don't care much how it's done. For Dan the aesthetics of the shot are vital, and when he shanks an approach, for instance, he is more disturbed at the ugliness of the shot than at being farther from the green than he was before he hit it.

On this Tuesday in March, Dan and I had to wait more than an hour after we had walked onto the El Rio course without a reservation. But then Dan proceeded to play a relaxed game where the bogeys did not rattle him and thus the pars and even the birdies came. On number nine, a long par five in which the trees narrow the target of each successive shot until the approach to the green is only a narrow window, I watched Dan study the landing area he wanted with each shot, watched him take a mid-iron instead of a wood for his second, and then watched him hit a precise wedge through the gap in the trees onto the green. He went on to shoot another seventy-nine.

10.

Mike Yots and Gerry Vidergar and I used to meet at the clubhouse at Randolph Park to go over the day's translation of the Old English epic poem *Beowulf* for our graduate class. Instead of each of us translating the two hundred lines for the next class, we would divide up the passage into sixty-five or so lines each, go over the whole thing quickly together, and if we were lucky, have time for nine holes of golf.

We usually started on the back nine, because we hadn't time to hit balls and number ten was straight, with a pretty wide fairway, and short, like most of the holes on the south course. (Randolph South has since been renamed Dell Urich, after the professional there for many years, the man who gave me my only golf lesson. All I remember of it was Urich's iteration, "Pronate! Pronate!" of the wrist move through the ball.)

Gerry's last name was sometimes a topic at these sessions; "gar" means "spear" in Old English, and though we could never find the first part of his name in our glossaries, we decided it must mean either "throw" or "shake", so out on the course, Gerry was always either "spear-chucker" or "Shakespeare". Of course the name wasn't going to show up in an Old English dictionary, being Slovenian; our threesome was, in fact, a poster group for Eastern European immigration to America: sons of Slovenians, Lithuanians, and Russian Jews.

11.

I was playing the crossover hole on the Old Course at St. Andrews. I had trained and bussed up, leaving

Katharine and two-year old Matt in Edinburgh on a murky, rainy morning. But just as I arrived at the Old Course, the sun had come out and there was an opening on the tee. It was June, 1972, and my first trip to England.

I hit my drive on number eleven, a 350-yard par four, into a fairway bunker. As I stepped into the sand, a woman strode briskly past me, hit my ball toward the seventh green, and strode on again, shouting back at me, "Keep op the pace of plee!"

12.

"Pretty good for an English major," said Wayne Bell, my colleague from the Math department, after I'd hit a nice drive on number twelve at Miller. The remark was ironic, of course, Math majors being even higher on the list of nerds and geeks than English majors.

A few minutes later the putt I needed for par came up an inch short.

"Another erg and it would have been in," Wayne said.

"Another erg and it would have been a jewel," I said, but then I had to explain the joke, not because Wayne didn't know perfectly well what a joule was, but because that reference was so unexpected from *an English major*.

13.

Then again, English majors can have their lapses, too. As we walked around the lake on "The Monster" — the 460-yard thirteenth at Miller, Richard pointed to the cattails and asked me, "What do you call those black birds with the red wings?"

Golf is a game that is helped rather than hindered

174

by bookishness. The aspiring young golfers in P. G. Wodehouse's comic short stories about golf always have their heads in books after the sun goes down on the course. Usually it is *Braid on the Grip* or *Vardon on Casual Water* or some other tome written by one of the game's early greats. I learned the fundamentals of the game from a book, Tommy Armour's *How to Play Your Best Golf All the Time*, in which the Silver Scot scolded me into agreeing with him that golf is a simple game, that the grip is its most important element, and that I needed to play the shot I knew I could execute and the one that would make the next one easy. Altogether this book is most notable for its gruff tone, as if Armour knows we're not going to take his advice. But it has helped my handicap get into the single digits.

14.

The fourteenth at El Rio is a wicked left-hand dog-leg par four, starting from the northwest corner of the course, with water as a danger to a pulled tee shot. Dan O'Neill and I sat in a cart waiting for the fairway to clear on December 17, 1992, the day after my twenty-fifth wedding anniversary. Dan was my wife Katharine's friend before I knew him; he had married Dee, a schoolmate of my wife's at Scripps Women's College.

We were sitting there reminiscing about more than twenty-five years of friendship, a lot of our time together spent on golf courses. He liked this course because he was a Tucsonan and this course was quintessential Tucson: it had been the original Tucson Country Club before the present course of that name was built, and it was the first venue for the Tucson Open, one of the oldest events on the PGA tour. It was a typical flat Arizona course, with tiny, domed Bermuda greens that were hard to hold and fairways defined by eucalyptus, salt cedar, and cottonwoods. From Tucson Country Club it had become El Rio (it was close to the Santa Cruz River) and would later be known as Trini Alvarez, for the surrounding park, the largest in the barrio that began in this part of town and extended to the south.

Dan reminded me that we had been playing this course on the day he got the word he had passed his bar exam. Then the golfers ahead of us moved on and we got up to hit our tee shots.

15.

Matt and Dan and I were playing Kiskiak Golf Course near Williamsburg, Virginia in October, 2000. On the fifteenth hole, Dan and I were watching an amazing comeback by Matt. He had gone four over on the first nine holes, but with none of the savage rage and self-blame that we remembered from his first years playing golf. Now he was even again, having birdied the par-four eleventh and twelfth and eagled the par-five fourteenth. But he was in danger of losing a stroke to par here at the fifteenth, having hit his approach iron well over the green. Matt looked carefully at the situation: the ball was sitting in a gulley filled with leaves, below the green, and the hole was forty feet or so onto the green but downhill from Matt's position. He struck the chip crisply with a short backswing and the ball came out fairly low, just carrying onto the putting surface. Then it first checked up before slowly trundling down to within a couple of feet of the cup.

"He's an alien," Dan said, shaking his head. "He's not from this planet."

"I learned how to do that from grandpa," Matt said. "I asked him one day how he always hit those chip shots so close and Herb said, 'I dunno. I try to keep my head down and just hit it.'"

I didn't quite get it. "Where's the big revelation there?" I asked.

"There wasn't one," Matt said. "That's just it. What he meant was a whole bunch of things: 'if you were watching closely enough, you could ask me something more specific I might be able to answer about that particular shot I just hit,' or, 'there isn't any way to tell you everything I do when I hit a chip shot without having gone through the routine of practicing and playing so that you know how the ball is going to bounce, how far it will run on this green, and so on.'"

Matt's game was about the same as mine and Dan's in the first years he played. In those days he would rage when he hit a bad shot, insist that his round was now ruined, and generally make an embarrassing fuss. But then, somehow he got his temper under control, and his game began to change. Because he has profound concentration — he can remember the layout and each shot he hit on a course he's played only once — he needs only reasonable practice and staying out of his own way to score well. His confidence, once he began playing well, was enough for anything: he has the conviction that if he knows how to play a shot, he can play it. Then he began to learn.

So when Herb said he just hit those chip shots, Matt understood. It was a Socratic moment with Herb as Socrates, denying his role as teacher, and insisting on the learner's having to make his way himself, with no short cuts, no magic tips. "It is a fine thing," wrote the fourteenth-century Japanese poet and essayist Yoshida Kenko, "when a man who thoroughly understands a subject is unwilling to open his mouth."

16.

Herb and I were playing the par-five sixteenth at Beatrice Country Club in August, 1995. This was the new Beatrice course, whose renovation was only about a year old, but the smooth greens and evenly-turfed fairways of the new holes already looked like a seasoned course. Just a year before, Matt and Herb and I couldn't play there because the construction was still under way. We played Hidden Acres instead, a course out in the corn fields on route two, where, on our second round, Matt had broken 80 for the first time.

The sixteenth at Beatrice is a 520-yard par five that plays out to the southwest corner of the course, and on this day it was playing against the wind. This is the new part of the course, one of seven holes cut out of farmland, with few trees but three small lakes added. High prairie grass borders the fairway, with only a narrow strip of short rough. Herb and I stayed out of the prairie grass, but it took us four shots to get on the difficult green, and we both three-putted for double bogeys.

It was my last full round with Herb, who stopped playing about a year later. In early 2000 he went into a nursing home, and he died a year later. That day he was visibly tired when we reached the seventeenth tee.

"I just don't have the energy I used to," he said. "I didn't expect to live this long."

17.

We were on Number Seventeen at San Ignacio. I was about to shoot an 80, a score that dogged me for a long time on those Green Valley courses, though I would occasionally break through and score in the 70s.

Matt was on the green and close to the hole on this par three. He would make birdie here and par eighteen for a 66. "I've never played with anyone who shot 66," Seth would say in wonder as we finished.

But it was Dan who made that hole memorable. After fifteen holes of badly hit woods from the tees, he'd decided he was going to finish the round using only a five-iron. His tee shot landed in the deep front bunker guarding this elevated green. Dan laid the blade of the five-iron wide open and hit a beautiful sand shot that landed two feet from the cup. Then he used the five-iron to putt the ball in for a par.

18.

My last round with Seth was in June of 2002, four months before he died. We played Haven on a day when the temperature reached 106°, but as we walked off the last green, he was still energetic and cheerful. Although he didn't score well that day, a few days before he'd played a round in the eighties, his best score of the summer.

He had called me the previous Thanksgiving to say he had been diagnosed with colon cancer. In surgery his doctors discovered a larger tumor than expected and nodes involved with other organs and so not removable. Like Herb, Seth had never expected to live so long, but he always thought his heart problems would kill him. Although he accepted his fate, saying he'd lived a good life and that no one should feel sorry for him, he agreed to a course of chemotherapy that may have slowed the cancer.

I managed to spend a week with him in early October, and later that month he ended our last telephone conversation, ten days before he died, by saying, "Love you forever."

17. The Place Where It Happened

On a March morning, recently, I was following the S-turns of Golden Gate Road, ascending its ridges and sliding a little on its soft sand as it dipped into washes. I was just over the Tucson Mountains in Saguaro National Park. In my early college years at the University of Arizona, my friends and I would speed along these roads—all of them dirt then except the road to Ajo, joy-riding in the middle of the night, hugging the inside of turns regardless of the possibility of oncoming traffic, sliding on sand and scree. We knew our young reflexes would save us from harm; in fact the practically deserted roads were the reason we survived.

This morning I was watching the roadside for flowers, but so far I'd been rewarded with only yellow brittlebush blossoms standing up from their low shrubs on taller stems and the pretty little stalks of orange flowers called globe mallow. Just as I was thinking that the paved roads would be better because of the run-off from the winter rains, a low hill opened on my left with patches of Mexican Poppies, Lupines, low Pygmy Daisies, white Desert Primrose, and many tiny wildflowers whose names I have never been able to remember as long as the next spring, even though I've dutifully looked them up in a guide. Suddenly a small burst of bright color erupted on my right, just over the little dirt berm created by the road grader when it passed. Even at the creeping pace I was driving, the color was past before I could get a good look, so I stopped at a wider spot in the road fifty yards on, put on my hazard blinkers, and walked back.

Plastic flowers surrounded a white cross upon whose horizontal piece, wider than its height, were the hand-printed letters of the name Dustin Jackson. The writer, who had printed the letters with a Magic Marker, had enclosed the name in quotation marks. A camo fatigue cap sat atop the cross, but most striking—and the note they struck for me was at first almost comic—were two boots angled out from the below the ends of the crosspiece. I wasn't sure whether they were Army-issue boots or biker's boots, but they were laced halfway up. A piece of what looked like a motorcycle's fender lay by itself on the dirt to the left.

After the first shock of those boots I could see well enough that this was a recent shrine or roadside memorial to a (probably) young man, perhaps in the military, who died recently as a result of a motorcycle accident on this spot. Growing up here in southern Arizona, visiting Mexico as I often did, I had seen many such memorials. This one was less elaborate than most, with no votive candles or mementoes left by friends or family. But a certain element of surprise always accompanies my coming upon these remembrances. I have been accustomed to ascribe it to culture shock: the shrines seem a part of grieving and remembering that is distinctively Latin and as common as Day of the Dead grave decoration. Thus another surprise awaited me when I got home that day and googled Dustin Jackson's name. He was thirty-six and a sergeant at Davis Monthan Air Force Base in Tucson, and he had recently returned from a tour of duty in Iraq—no surprises there. But Jackson was from Red Oak, Iowa, unmarried, and I looked in vain for a Hispanic name among his surviving family.

A little more googling brought me some general information about roadside shrines. Sylvia Grider, an anthropologist who has studied them, wrote in a *New York Times* article in 2009 that the custom of placing such memorials has spread from the American southwest to the entire country. Apparently you don't have to be Hispanic to want to memorialize a dead family member or other loved one with this sort of shrine. But not everyone thinks it is a good idea to do so. In the blog that followed the *New York Times* article, one mother of a car accident victim wrote, "Grief is a private process. It needs to stay private." And another woman wrote, "I find them macabre."

One can't argue with an affective reaction like the second writer's. But the conviction of the first woman made me wonder. Does she mean that everyone's grief needs to stay private? Or is she simply asking, in another way, the question that used to occur to me: why do the shrine-makers feel a need to take their grief to the place where it happened?

***

A few miles outside of Tucson on the road that leads

to the town of Ajo and its huge but now inactive open-pit copper mine is another memorial to a motorcycle rider. Much more elaborate than Dustin Jackson's, this one has an actual altar made of bricks, upon which are placed plastic flowers in vases, votive candles, a large bear fetish crudely carved from flagstone, two tiny orange traffic cones, various tokens of or models of motorcycles, and a three-foot high plaster angel festooned with flowers, rosary beads, crosses, and tiny motorcycle charms. A cross stands to the side of the altar with the name Donald spelled out in small letter tiles, the name of his motorcycle club, Southside M. C., and the years of his life, 1957-2004. A stanchion behind the altar holds up a small, solar powered garden light with a plastic Blue Jay wired to it, but what is most conspicuous is the motorcycle wheel supported two feet from the ground on a metal standard, with shining chrome spokes and a bright reflective hub. The rim of the wheel is bent out of round at the bottom, and black scuff marks mar the top of the whitewall trim. An aluminum strip around the standard identifies the motorcycle as a Harley-Davidson Heritage Softail.

Donald Borquez died here on D-Day in 2004, sixty years after the Normandy invasion but only forty-seven years after his birth. His memorial sits well back from the road, almost to an irrigation ditch that parallels Ajo Road, but that shining wheel in the air is readily visible as one approaches from either direction. The clearance from the road is an important issue in Tucson's regulation of roadside shrines. The state of Arizona has historically been permissive about the matter. Years ago when I was growing up here, the highway patrol itself set simple white wooden crosses at the sites of road fatalities. Sometime after I moved away to a teaching career in other states, the practice was discontinued, but families of accident victims then picked it up, placing either the crosses or more elaborate memorials at the sites. Tucson's regulation states simply that roadside memorials "may be left in place within the City of Tucson's rights-of-way as long as they are well maintained by others and do not pose a safety hazard or sight visibility issue."

***

From a distance you might mistake Kevin Robinson-Barajas' memorial for one of the "ghost bikes" that are sometimes seen in American cities and elsewhere in the world. According to the website ghostbikes.org, cyclists who die on city streets are sometimes commemorated by friends who paint a bike all white and lock it "to a street sign near the crash site, accompanied by a small plaque." There are now hundreds of these memorials throughout the world. But the ghost bikes have more than a personal remembrance for their purpose; they are reminders also of the danger cyclists always face in city streets. Though Kevin's memorial is at a busy Tucson intersection, it clearly is much more personal. And, as I stop at the Circle-K across the street and walk nearer, I can see that Kevin's bicycle is not white but blue, and festooned with plastic flowers and a tinsel wreath.

Kevin's bike remains upright because its back wheel is encased in Quikrete. On the leveled surface of the concrete is a stylized, nearly life-sized Gila Monster. The bike stands between standards that hold up a street light at one end and the traffic signals at the other. The intersection is that of Fort Lowell Road and Mountain Avenue, where a school bus struck and killed Kevin in September of 2008.

Near the back of Kevin's bicycle is a small group of

184

rocks, some quite small and others as large as two feet on a side. A cross, vases of flowers, and votive candles are placed among the rocks. The blue and green hues of azurite and chrysocolla gleam from some of the rocks, obviously from a copper mine somewhere in the area. Others shine white in the sun, and when I look closer

I see these are mostly quartz. Perhaps the rocks and the stylized lizard point to an interest in nature in the boy. One of the larger stones seems to be granite, not dressed or smoothed but engraved deeply on one irregular surface with these words:

Remember
Kevin Robinson-Barajas
age 15

***

This straightforward appeal to us may be the essence of this and other roadside shrines. They are conspicuous sites of the grief of families, certainly, but they also speak directly to the passersby and remind us to remember Kevin and remember, too, that we will also die. The reminder is seen by some as morbid; this word comes up often in the reactions of the shrines' critics.

"Why doesn't a memorial in a cemetery suffice?" is a question often asked by critics of the shrines. Those who make them have several responses to this question: a typical one is "this is where my loved one's spirit was last on earth." Another response is that the roadside site is in the land of the living, a place where the mourners can think of the dead loved one as actually having been, while the cemetery holds only remains, is a place of the dead, not a place in the world where he or she lived. These feelings are mixed, however, with the painful ones of remembering that the memorial marks where a death occurred.

Other critics of the memorials have trouble with such a public display of grief. "But, sure, the bravery of his grief did put me / Into a towering passion", Hamlet says of Laertes's extravagant show of mourning at Ophelia's grave, and Hamlet is only one of many who

186

are less than tolerant at such a show. Still other critics believe their objections are more logical. The 2009 *New York Times* article I mentioned also contains an attack on roadside shrines by a lawyer who some years ago successfully defended a client who had been sued for removing a roadside shrine. (Though many states tolerate the shrines, so far as I know only one, New Mexico, has a law protecting them.) The lawyer cited several reasons for not tolerating the shrines: in his view their presence amounts to a taking of public land for private purposes; he sees crosses, rosaries, and other religious tokens that are often part of the shrines as a violation of church-state separation; and he believes the shrines constitute a distraction and a hazard to drivers. Each of his points is contested many times by respondents to the *Times* blog.

***

On a July morning in 1998, my wife and I were driving along Interstate 40 toward Johnson City, Tennessee. Very early that morning we had been awakened by a telephone call from a hospital in Johnson City. During the night our younger son Dan had been involved in an accident on the stretch of I-40 we were now driving. He had come upon an accident that had just happened. Pulling over to the side and turning on his hazard lights, he walked toward a car without lights, stopped on the roadway, partially blocking the passing lane. He moved around to the driver's side and was asking the man behind the wheel if he could help when a semi-trailer truck, moving away from the hazard lights on the right shoulder toward the inner lane, struck the stopped car on the passenger side, throwing Dan thirty feet into the highway median.

Miraculously, he was not killed. In fact no one died in either the first accident Dan had stopped for or in the truck's crashing into the already damaged car blocking the passing lane. Dan escaped without even a broken bone, though he was bruised all over and was unconscious for a brief time, waking up in the helicopter MedEvacing him to the Johnson City hospital. But he had no concussion, no fractures, and no internal injuries. He had called us from the emergency room after he had been thoroughly examined.

We were driving now to take him clothes—his had been cut off in the emergency room—and to take him to the auto yard where his car had been towed, now with a dead battery after its hazard lights blinked for many hours. As we drove we looked carefully along the road for any sign of the two accidents that had occurred somewhere there, in rapid succession, during the night. We could see no sign—no broken glass, no pieces of bumper, not even skid marks. We continued scanning the roadside until we were in the outskirts of Johnson City. And then, simultaneously, we both had the same thought. My wife gave voice to it: "If Dan had died there, we would know by now exactly where it happened. And we would never forget that spot of highway."

18. A Retiree Reads Proust and Montaigne

We must reserve a back shop all our own, entirely free, in which
to establish our real liberty and our principal retreat and solitude.

> – Montaigne, "Of Solitude" (1588)

My friend Hughie Lawson, having retired from teaching history at the college where I also taught, decided the first thing he wanted to do when his time was
wholly his own was to read Proust. And read him
he has, twice traversing *A la recherche du temps perdu*
with a good French dictionary and only the occasional
help of a crib—the excellent translation begun by C.
K. Scott-Moncrieff and carried on by Terence Kilmartin
and Andreas Mayor. Now Hughie, still as fervent as
any late convert, is listening to tapes of the first volume, *Du côté de chez Swann*, and he has made another
convert in me.

At Hughie's urging, I determined to read *Remembrance of Things Past* once I had joined him in the leisure
of retirement. *Swann's Way* had been assigned in my
sophomore humanities course, but all I could remember of it 40 years later was the pet name Odette and
Swann used for their lovemaking, "doing a cattleya"
(because their dalliance began when he was adjusting a cattleya orchid on her breast one evening). I had
never even looked into the other six parts making up
the three-thousand-page novel. As I began rereading
Swann's Way, a couple of thoughts struck me. I wondered what in the world I had made of this book when
I read it at twenty. And I became captured by the bizarre, hypersensitive consciousness of the narrator
("neurasthenic" is the word he uses for his condition).

I was entranced at the narrator's deliberate and precise anatomization of his own mental and emotional life. He records his aesthetic development as well: the writer Bergotte instructs him in reading literature, the dilettante Swann gives him confidence in every social milieu, and the painter Elstir shows him how to see the world. The narrator describes a rich and large cast of characters — the old family servant, the self-important diplomat, the pedant, the doctor, the duchess, the musician, the soldier, the tailor — and distinguishes each of their voices. In my experience, only Jane Austen is as accurate a mimic in producing dialogue where I can instantly recognize the speaker.

I was also reading Montaigne at the time. My writing projects changed from academic articles and books to personal essays when I retired, so it was natural for me to become absorbed in reading the man who invented the personal essay. Where Proust was a novelty, Montaigne was an old friend who gave me the pleasures of familiarity. Montaigne does not enforce a linear approach; I can pick him up at the beginning of any of the more than a hundred essays. With him I can fall into my habit with any collection: I read the shortest piece first, then the next shortest, and so on. Or if I have the inclination, I can begin with a longer essay. "If you will not give a man a single hour," writes Montaigne, "you will not give him anything." And what he gives in return is always worth the effort. On one spread of two pages he takes me from Sejanus to King Clovis of France to Mohammed II to Antigonus the Macedonian and on to the Russian Jaropelc and the king of Poland. But he is never far from his main subject, which is Montaigne himself.

Montaigne is an expert on retirement. He knows that in the absence of pressing public matters, important duties and tasks, the small domestic ones assume undeserved importance. And he knows the vital importance of holding back a part of oneself. "We have lived long enough for others; let us live at least this remaining bit of life for ourselves." "Let us keep," he urges, "a back shop (*une arrière-boutique*) all our own, entirely free."

Both these writers provided me with what I was after: some self-indulgence and a good dose of the subjective. Proust's almost incredibly self-indulgent narrator was a profound liberating influence on me, stuck as I had been for years in academic writing where the "I" was an irrelevance. I had written about English literature and English painting, about Shakespeare, about reading poetry, and about mystery fiction, but outside of prefaces I never got to write about how or why these things moved me personally. I wanted to write about myself for a change. Proust implies that no other subject exists. Montaigne's concentration on himself is less airlessly solipsistic, but even more encouraging for a writer in ways it took me some time to appreciate.

Proust and Montaigne, two French writers working three and a half centuries apart, both assembling one major work by accretion, then tweaking, adjusting, adding and modifying, both have themselves as subject and time as theme. Each convinces me that he has all the leisure he needs to spin out what he takes up, to see where it goes. (Proust creates this impression even as he is racing against death to finish his book, and losing the race.) Each often reminds me of a naturalist who sets out on a day of bird watching but spends most of the morning examining with a magnifying glass a potato beetle that happened to catch his attention. Each writer seems to move away from his main structure and back to it, without apology and without regret, assuming I will follow. They have also in common an attention to the process of writing.

192

All three of us, Proust, Montaigne, and I, have retirement in common. "Let's face it," a blunt friend of mine said, rather unkindly I thought, "the narrative of your life is over. Now you're trying to make sense of it." Montaigne retired to his country estate to write and revise essays. Proust retired to a cork-lined, windowless bedroom to write and revise his huge novel. They're both experts at "making sense of it" by writing.

***

Two writers of the same language could hardly appear more stylistically different. A purist might even argue that the early twentieth-century modern Parisian French of Proust and the late sixteenth-century Gascon-sprinkled French of Montaigne are not even the same language. The most striking feature of Proust's style is the sentence structure, with many long relative clauses and parentheticals. My friend Hughie says the clauses "seem like one of those puzzle boxes one inside the other", but he is convinced this complexity has to do with Proust's theme: "One of my wacko theories," he writes, "is that since he was writing about memory he built this into the grammatical structure by writing such long sentences that you have to work to remember in the middle of one how it started out." But, my friend adds, "Proust is always fair with you, and his sentences are always grammatically proper, so that in principle each could be outlined, the way Mrs. Sheram taught us in high school, but the average one would require the whole chalkboard!" I know that Hughie is right about this, because I have worked through those sentences until I saw the link between every modifier and its antecedent, and yet, I also find an apprehension mounting to a small panic when, having been reading a sentence for a hundred words or so, I glance down the page and then on to the facing page and find no paragraph break, no full stop even, and my suspicion becomes more certain that Proust is going to keep pursuing the point that he began with and has since narrowed to a still finer one, revolving it in the light, identifying its colors and then counting the nuances of each shade.

Montaigne moves along quite rapidly by contrast, and his style is relatively spare and concise. He tells us that one of his models was Tacitus, whose *Histories* of Rome Montaigne read in one sitting — very unusual for a man who tells us he rarely spent more than an hour at a time reading. Montaigne avoids the extreme conciseness and the sometimes monotonous oppositions of Tacitus because he was worried about being "too compact, disorderly, abrupt, individual." He even makes fun of these aspects of the Roman historian's style by parodying them: "The unrolling of public events depends more on the guiding hand of Fortune: that of private ones, on our own. Tacitus' work is more a judgment on historical events than a narration of them. There are more precepts than accounts. It is not a book to be read but one to be studied and learnt."

Plutarch and Seneca were also favorites. From these writers he learned to seek clarity and straightforwardness; he quotes Seneca's rejection of stylistic elegance as *unmanly*, a revealing word. Two hundred years before Buffon's observation that "Le style est l'homme même", Montaigne argued the connection between how one writes and how one lives one's life. To write well and simply is an ethical, specifically a Socratic, imperative. Lack of ornament is one reason that rarely does a Montaigne sentence stretch past forty words.

Of course he digresses. Reading Montaigne I sometimes have a diluted version of the same panic Proust inspires, that the writer could pursue this particular line of elaboration or these parallel examples from Roman history for much longer than can reasonably be imagined or forborne. But then I relax, remember that I have retired, and smile slyly to myself. Montaigne, also a retiree, away from public duties at his estate, can pursue his thoughts at leisure. I have the leisure to pursue him as he pursues them. When Montaigne really strays from the mark (he reminds me that sometimes when I think he is digressing I have merely failed to see what his point really is), it is not necessarily in one of the longer essays. The essays in the third book almost never strike me as distractingly digressive, while some of the earlier, shorter essays can do so; in these I often think Montaigne is ranging afield to avoid talking about himself. In the later essays he relaxes into more extensive self-revelation, and the pertinence of the examples becomes clearer as the real subject stands away from the background.

I am captivated by the humor of these two writers and in the disarming self-deprecation with which Proust's narrator and Montaigne's "I" present themselves. Montaigne is short, he tells us, "an ugly defect" that embarrasses him when strangers look over his head, mistaking him for a servant and looking around for his master. He prefers riding to walking because "in our streets small men are subject to being jostled and elbowed, for want of presence." Proust's narrator is clumsy and gets caught in revolving doors. He is frequently off-balance in some social way: foiled in trying to get an introduction to the Duchess and later to the Prince de Guermantes, not quite sure his invitation to the Princess de Guermantes' reception isn't a hoax, vainly and comically trying to meet the supposedly sexy maid of the Baroness Putbus, on the wrong side of the woman he's supposed to take in to dinner so that she has to pirouette to his right side to hide his gaffe. Montaigne accuses himself of clumsiness, and he adds having a bad memory and being a poor storyteller. "Not so," is my response, nor can I go along with him when he calls his essays "some excrements of an aged mind."

Along with self-deprecation, Proust and Montaigne share a satiric view of the inability of those around them to look at themselves and the world with a long view. Thus fashion, for example, attracts their scorn. Montaigne is as devastating on codpieces as Proust on monocles — in more than one place Proust talks about men wearing monocles being like fish carrying around a little piece of the aquarium which they stare out of. Montaigne has little patience with religious fanaticism and writes that "it is putting a very high price on one's conjectures to have a man roasted alive because of them." He makes fun of world travelers who come home having learned only "the measurements of the Santa Rotunda, or of the richness of Signora Livia's drawers." Proust anatomizes the cynicism that could be completely aware of Dreyfus' innocence yet keep him jailed for years.

Both can be satiric about doctors and medicine, and both observe that the attendance of doctors prolongs naturally short illnesses. Montaigne says that doctors get to bury their mistakes while their successes walk around as advertisements, that they get the credit for cures effected by nature or anything else aside from their medicine, and that those under a doctor's care seem to be constantly sick. In Proust, doctors are likely to be shown as ignorant, obsessive, and unable to put themselves in others' places. But characteristically, both writers stop short of extremism here and elsewhere. Proust's narrator warns us that dismissing doctors and medicine would be a mistake: "to believe in medicine would be the height of folly, if not to believe in it were not a greater folly still." This caveat comes from one of the deepest convictions that the narrator's poor health has instilled in him: the body will not be ignored. Both Proust and Montaigne deplore the behavior of doctors, but both also recognize that we all conspire to create it: because what we expect of doctors is so unrealistic in the long view, we can hardly fault them for pretending to give it to us. Those of us who aspire to the long view, if we have managed to attain any of it, have inevitably some years upon us; doctors and the body are naturally part of the prospect.

The ineluctable body is a major theme for both Proust and Montaigne, and neither has worked out the puzzle of the relation of body and mind. On the one hand, the body's demand for attention in sickness (and age) affirms the dualism of the self: my thoughts can be healthy while my body is sick. "It is in sickness," writes Proust's narrator, "that we are compelled to realize that we do not live alone but are chained to a being from a different realm, from whom we are worlds apart, who has no knowledge of us and by whom it is impossible to make ourselves understood: our body." On the other hand, a condition of health is the integration of mind and body. "The body has a great part in our being," writes Montaigne, "those who want to split up our two principal parts and sequester them from each other are wrong."

***

At first I found the hourly experience of reading *Remembrance of Things Past* much like lived life in its pace and seeming formlessness, but as I spent weeks and months with the book I began to see Proust's careful attention to form and cohesive structure.

To pick one example, the old music-master and composer named Vinteuil is used to link the three great passions of the book. Charles Swann hears a phrase from a Vinteuil composition for the first time when he meets the courtesan Odette de Crécy, who will be the love of his life. The musical phrase becomes the theme for Swann's infatuation. In the middle of the book, the narrator's chance mention of Vinteuil's name leads Albertine, the woman whom he is about to dump, to reveal her lesbian passion for Vinteuil's daughter. This revelation, because of the economics of love in the novel (the unavailable is precisely the longed-for) cancels the narrator's decision to break with her and puts him in the same position Swann was when besotted with love for Odette. Near the end of the novel, an entire Vinteuil septet (which contains the phrase that so enchanted Swann) becomes in turn the theme for the homosexual passion of the Baron de Charlus for a young musician.

And so it went as I proceeded through my seven-months' reading of *Remembrance of Things Past*, whose accumulated and connected or parallel episodes became *my* memories, reawakened by later happenings, just as they are for the narrator. To build this structure Proust wrote three versions of *Remembrance of Things Past*. When he was in his twenties he wrote an early outline of the novel as a third-person narrative. This manuscript was published after his death and given the title of its main character, *Jean Santeuil*. It does not take the protagonist very far in his progress toward becoming a writer. Ten years into the new century Proust completely rewrote the novel, enlarging it into three volumes. The first was published as *Swann's Way* in 1913, and presumably the other two volumes would have followed over the next two years, but the war intervened, and during the war years Proust began to expand the three volumes into what would eventually span seven. The second volume, *Within a Budding*

Grove (A l'ombre des jeune filles en fleur — sometimes Scott Moncrieff's English translations of the titles differ from Proust's titles), won the highest French literary honor, the Goncourt Prize, when it was published in 1919. *The Guermantes Way* came out in two parts in 1920 and 1921; *Cities of the Plain (Sodome et Gomorrhe)* was also published in two parts in 1921 and 1922. After Proust died late in 1922, two more volumes that he had completed were published, *The Captive* in 1923 and *The Sweet Cheat Gone (Albertine disparue)* in 1925. The last volume, *Time Regained*, was clearly not ready for publication when Proust died, for it contains many inconsistencies, but the bulk of it was complete and it was published in 1927. The edition I read, based on the Pléiade edition of 1954, is the work of editors who collated the many changes — chiefly additions — that Proust made to the whole text right up to the night of his death.

Montaigne's *Essays* also result from a three-stage process, which may be thought of most conveniently in terms of its three books or volumes, but which editorially is much more complex. Where Proust's stages are metamorphic, Montaigne's are accretive. Montaigne began writing essays after retiring to his estate in the Bordeaux region in 1571. He published two books of these — 57 and 37 essays respectively — together in 1580. Subsequent editions of the first two books, with many changes and added passages, came out in 1582 and 1587. Montaigne added a third book of 13 essays in 1588, and until his death in 1592 he continued to revise and add hundreds of passages in a 1588 copy of the *Essays*. He says himself of the growth of the text, "I do not correct my first imaginings by my second — well, yes, perhaps a word or two, but only to vary, not to delete." In all, the 107 essays are only about a third as long as *Remembrance of Things Past*, but still hefty at around a thousand pages. Several essays are less than two pages, while the longest, "An Apology for Raymond Sebond", is over 150 pages.

Many of the first series of Montaigne's essays are tentative and diffident, stingy in their self-revelation. Frequently the first-person pronoun will appear only once or twice. Anecdotes from classical sources and from European history make up the bulk of these essays. Gradually the pattern changes. In the second and third series, the essays begin to lengthen. "Because such frequent breaks into chapters as I used at the beginning seemed to me to disrupt and dissolve attention," Montaigne writes, "I have begun making them longer, requiring fixed purpose and assigned leisure." It's as if, in mentioning these two requirements, he were writing directly to me, and I begin to discover more and more about Montaigne's early education, about his father, about his friend La Boétie's last days, about the onset of Montaigne's kidney stones, about his habits in travel, eating, sleeping. The allusions to history and contemporary events become fewer, and serve other purposes. Instead of parallel instances (both Proust and Montaigne operate by juxtaposing examples and parallels) bolstering Montaigne's opinion, or even hiding it, the examples begin to demonstrate that the world is a larger place than the reader imagines, with more diversity of customs and ways of looking at experience.

China, Montaigne writes, is a kingdom "whose history teaches me how much ampler and more varied the world is than either the ancients or we ourselves understand." The later examples strike me as smaller instances of the way the whole essay "Of Cannibals" works. In that famous essay Montaigne starts with an anecdote — really three parallel anecdotes — of Greeks who, though taught to consider Romans barbarians, when they actually encountered Romans observed that their behavior was far from barbaric. The moral is that we should judge from observation and reason rather than vulgar report, and the method is anecdotal and empirical. The essay as a tentative form must find its own structure, constructing unity and connections; here the Greek looking at the Roman parallels the Frenchman looking at the Brazilian native. And the ending shows the search for form as well: where the storyteller has an active resolution to end his narrative and the treatise-writer exhausts analysis of the subject, the essay writer must decide where to stop and convince the reader that it is an ending. Montaigne ends "Of Cannibals" by saying he had a long conversation with one of the Brazilians who had been brought to Europe, but the incompetent interpreter did not understand Montaigne's questions. "But what's the use?" he concludes, "They don't wear breeches." The ending comes back around to the essay's assertion "that each man calls barbarism whatever is not his own practice" and suggests the difficulties of any encounter with otherness.

When Montaigne describes the natives' custom of dispatching and eating their prisoners, he does not excuse them but says seeing their faults should make us see our own. In Montaigne's hands the essay form seems eminently suited for this kind of reflexivity: when it seems to range farthest afield, it comes back to an examination of himself or his culture.

***

Because I write, I find Proust and Montaigne especially interesting when they reveal how their works came to be written; that is, what both enabled and forced them to write and why the writing took the form it did. In Proust, I found three passages very revelatory in this respect, one at the beginning, one in the novel's middle, and one near the end. The whole of the book traces the progress of a young man toward the realization of his vocation as writer.

The first of these passages, in the section titled "Combray", is the most famous piece of writing in the novel: where the taste of a little cake called a madeleine dipped in tea calls up for the narrator a complete memory, replete with smells, tastes, sights, sounds and textures, of a Sunday morning from his boyhood, and in so doing fills him with a feeling not merely of joy, but of a mysterious freedom from contingency and mortality. I found the scene so carefully realized in its sensory detail that I did not realize what significance it was going to have until later in the book. Here, long before he intimates what the role of art might be in his life, or the role of memory in art, the narrator makes an important point about how memory constitutes the self.

The second passage, at the beginning of *Cities of the Plain*, and often referred to as "La Race maudite", was a revelation for me rather than for the narrator. Here I suddenly realized why Proust prefers the term *inversion* to *homosexuality*. The narrator, who is a young heterosexual male, happens to see a scene in which the Baron de Charlus and a shopkeeper recognize each other's homosexual attraction for each other. In describing the Baron de Charlus, Proust looks at his own experience of homosexuality from the outside, from the straight point of view. The narrator details the maneuvers, the reversals, the switches, the *transpositions* and *inversions* the Baron and those like him are obliged to perform "to make a secret of their lives" and how they must "change the gender of many of the adjectives in their vocabulary." I imagined myself inside Proust's mind, looking at homosexual experience from the straight point of view and seeing how he could write about it by transposing it to look like heterosexual experience, separating experience from language. Proust gambles that the depiction of the lover's projection of his image of love onto the beloved, the growth of love, and the jealousy of the betrayed lover will all resonate with the gay or straight reader alike. The transposition works as a metaphor for all the transformations the writer must make in separating language from experience. The author's experience is viewed from the outside and transformed to become the narrator's experience, so that Proust may say truthfully, if somewhat coyly, as he does late in the novel, that "there is not a single incident" in the book "which is not fictitious, not a single character who is a real person in disguise."

Homosexuality as a perspective not only connects sex and art in *Remembrance of Things Past*; it links its whole cast of characters, from the Faubourg Saint-Germain (the old aristocratic quarter of Paris and Proust's blanket term for high society) to the lowest climbers on the rungs of society's jungle gym. One of Proust's topics is the way society changed during his lifetime, leveling the aristocracy, mixing the classes, and allowing the influx of *hoi polloi* into previously aristocratic playgrounds and watering places like his invented Balbec. Proust was able to see these changes from the top, though himself a bourgeois, because he was a regular guest at parties of the noble and the near-royal. No doubt it was easier for him to transcend class limitations in part because he was a member of that club (or "freemasonry" as the narrator puts it) where the barons and the bourgeoisie associate in "perilous intimacy".

The last key passage comes near the end, when Proust's narrator steps on uneven paving stones that recall Venice for him in the complete sensory way that the madeleine recalls Combray early in the book. He suddenly realizes that he *is* a writer, after doubting it for many years. There is some mystification in the narrator's account of the creative process here, but it is clear that he decides the role of the writer is to conquer time by somehow allowing art and memory to make connections that are made only by physical experience in the ordinary world.

The genesis of Montaigne's essays, if we believe the author, was much simpler. He talks about it in the early essay "Of Solitude", in the late essay "Of Vanity", and in other essays briefly. He began to write essays because he recognized, having retired to his country estate, that his talent was neither for household management nor for bookish study. As Virginia Woolf says, he *needed* to write, because writing for him was health, truth, and happiness. Writing was pleasure-giving but also necessary. "This bundle of so many disparate pieces," he says, "I set my hand to... only when pressed by too unnerving idleness." That his essays took the form they did was largely a matter of finding or inventing a form. Montaigne tells us that his correspondents thought him quite a good letter writer, and that if at the time he came to write the *Essays* his friend Etienne La Boétie had still been alive, we would have the *Essays* in epistolary form. Beyond these comments Montaigne does not go much farther. His comment about "unnerving idleness" suggests that writing was a hygienic activity for him. Writing never assumes for him the obsessiveness it has for Proust at the end of the latter's life. Yet he writes the essays and rewrites them, for the last twenty-odd years of his life.

Proust and Montaigne appeal to me because they are such conscious writers — writers' writers — and part of this is their demonstration of the ways in which they come to writing. But each is also concerned with writerly matters such as the relation (and proportion) of narrative to reflection, the mode of address to the reader, the degree of authorly intrusion, and the mechanics of getting on with it. "Let us here steal room for a story," Montaigne writes disingenuously (where was he taking me at such a determined gallop that this momentary drawing in of the reins will be such a shock to me?). He thus draws attention to the way of proceeding and the enterprise from which room must be "stolen" to tell a story. Proust's narrator would be as disingenuous were he to writes, "let me steal room from my story to tell you what I thought about it at the time and what I think about it now." Proust's "story" is as much commentary as Montaigne's commentary is story. I especially appreciated an "interruption" in the middle of *Remembrance of Things Past* where the author has a conversation with a nettled reader, who complains about digressions, raises the question whether the narrator is in fact the author, asks what happened next, and is rebuked ("be quiet and let me go on with my story"). Such give and take is not unusual in either writer, and I find Montaigne chivvying the reader as well: "It is the inattentive reader who loses my subject, not I," he says of the implied criticism that he doesn't stay on the topic indicated by the title.

***

For both Proust and Montaigne, the important life, the one that is most real, is the internal. For both, writing the internal life functions as a way of keeping them human and saving them from the eccentric or the monstrous. In writing, they construct themselves: each describes the development of a self, an identity. Proust does it in narrative that takes his character from early boyhood to adulthood and then to the threshold of a career as a writer. Montaigne does it in the way his prose gradually drops away the classical props and the authoritative quotes; he grows into the assurance that he is the only authority on his subject.

The subject in each case is the writer, or rather the character who speaks to me in first person as if he were identical with the writer. I am wary of taking the one for the other in my reading, but, oddly, with these two writers it does not seem to matter; whether or not the author has constructed a persona for himself in the character who calls himself "I" or not, the profound exploration of subjectivity goes on. Of course I tend to read the *Essays* as something much closer to autobiography than I would ever consider *Remembrance of Things Past*. And Proust lets me know that I need a more subtle approach than reading the novel as a key to events or people in his life: his narrator's coyness about his name ("If we give the narrator the same name as the author of this book, [it] would be... Marcel") and his denial, late in the book, of any one-for-one correspondence between characters in the book and real persons — these underline for me what I knew already: that the narrator is not Marcel Proust in any circumstantial way but *is* in every important one.

Montaigne, on the other hand, signals in every way that he is indeed writing about himself circumstantially. He uses his name, which is also the name of his estate and of his father. He refers to events in France and in his region, always with the assurance that I will accept their truth. "You remember those terrible days of the religious wars," he seems to say, or, if you do not, "you could look them up."

So with each the content is the sum total of things that happened to their subjects. Proust's subject is in love early and often, but finds himself alone at the book's end. Montaigne's character was licentious in his youth, he tells us, and although he marries and has a daughter whom he loves, he also writes as if he were alone in the later essays. Each man lives through a turbulent time in France. Proust's narrator is clearly a Dreyfusard in the early period when such a position was unpopular, but he lives to see Dreyfus' innocence largely accepted among all classes of French society. Later, he spends the war years in a sanatorium. Montaigne lives through a time of religious and civil unrest when his own moderate position is a problem: "to the Ghibelline I was a Guelph," he writes, alluding to the disputes in Florence during Dante's time, "to the Guelph a Ghibelline."

***

I enjoy being in the company of these two writers, because they provide pleasure and because I find a special message in them for me in my present stage of life. They have other shared qualities besides their humor and self-deprecation to make their company enjoyable: both believe in the Socratic approach to gaining knowledge, both are moralists first and always, both are tolerant, both believe in the self as an accumulation of memories and habits. Neither shrinks from the dark side: Proust painfully anatomizes jealousy; both writers talk about the indifference with which we can watch the sufferings of others — in fact both quote the same Lucretius poem beginning "suave mari magno" about such indifference. Both are very concerned with the craft of writing, attentive to the relation between reader and writer, not above chiding the reader for lapses of attention or failures of confidence in the writer; each is committed to a huge project he sees formally as a single entity and determined to put everything he knows into it, dying in the attempt.

Stuart Hampshire, who introduces Donald Frame's translation of Montaigne, spends a couple of paragraphs of contrast between Montaigne and Proust. He makes much of differences he thinks essential: Proust has a "message" about art and imagination; Montaigne is "interested only in his own subjectivity." Proust is a pessimist where Montaigne is not. In my view, Proust's "message" is as much about morality as about art, and Montaigne's subjectivity is always the moral self-examining itself. Whether it makes sense to talk about pessimism or optimism I am not sure: Montaigne speaks of his cheerfulness, it is true, and Proust does not seem to hold out happy prospects for human love. But in the larger sense they ring the same tune, and it is the experienced observer's judgment: *sumus quid sumus*, we are what we are.

No novel writer can ignore Proust. He represents an ultimate fulfillment of the Richardsonian tradition of the novel, where emotional nuance always trumps cruder forms of action and the limiting case is a complete transcription of the narrator's mental processes, so detailed that reading time exceeds the lived time being described. Proust approaches that limit so closely that he scares away timid writers working in the tradition. Even Virginia Woolf writes to Roger Fry, "Oh, if I could write like that ! I cry," and confides to her diary about *Mrs Dalloway* "I wonder if I have achieved something. Well, nothing anyhow compared with Proust... and he will, I suppose, both influence me & make me out of temper with every sentence of my own." But I am not a novelist, and Proust doesn't scare me. For me, reading Proust is proof of my real liberty to choose where I spend my time. Seven months I spent in his company, and the enjoyment was partly in the choosing: shall I stop after *Swann's Way*? After two parts? Halfway through? I enjoyed the reading as I enjoy any novel, but there is an additional pleasure for a writer — any writer — in Proust's tour de force performance.

Reading Montaigne, in addition to its obvious pleasures, has on me as a writer an effect nearly opposite

that of the usual anxiety of influence: he makes me feel I
can do it. Lewis Thomas also thinks we generally come
away from Montaigne with encouragement: "He is, as
he says everywhere, an ordinary man," writes Lewis,
but "if Montaigne is an ordinary man, then what an
encouragement, what a piece of work is, after all, an
ordinary man!" Not only does Montaigne avoid pre-
scribing form—that is, he works in a form that he describes in tentative terms, as an *essay* or merely a trial
or attempt—but he also avoids dramatizing himself as
Author. The self with Montaigne is not an idea, as it is
with Whitman, for example, but an unabstract thing, a
real and vulnerable thing that pisses black when he has
the stone. He disintegrates the authority of the self by
writing about the disintegration of the body. He vali-
dates the attempt, the trial, the effort. From the four-
hundred-year-old Frenchman to the gray-headed retired English teacher comes a benediction: retirement *is*
that back shop, that *arrière-boutique* we reserve to ourselves, "in which to establish our real liberty and our
principal retreat and solitude."

19. On Not Being E. B. White

1.

When I was in college I roomed with the family of a friend one summer. My friend's dad was an old Scot who had fought in WWI and married late in life. He would stick his white head into our bedrooms in the morning and always repeat the same verse:

> Up lad! Thews that lie and cumber
> Sunlit pallets never thrive.
> Morns abed and daylight slumber
> Were not meant for man alive.

Well, I knew it wasn't a Scots poet, although the lines sound right to me in memory only when inflected with Alan Mail's gentle northern burr. I knew it was an A. E. Housman poem from somewhere in *A Shropshire Lad*. And though I never bothered to look the poem up, that one verse stayed in my memory. When I had my own boys I sometimes woke them with the Housman lines. Then one early summer morning, when my older son, Matt, had just returned from his first year at Oberlin College, I opened his door and repeated the Housman verse at him as he lay, facing away from me. He turned over and, opening one eye, capped my verse in a sleepy voice:

> Clay lies still, but blood's a rover,
> Breath's a ware that will not keep.
> Up lad! When the journey's over
> There'll be time enough for sleep.

My own blood stopped roving for a few seconds. His quick reaction, coming so unexpectedly when I'd thought him at best half awake, was a shock, but then the sense of the lines hit me and chilled me. I might have known what the real subject of the poem would be, since it was by Housman, after all, with his too-soon-dead light foot lads and rose-lipped maidens. I should have known that even in the one stanza I knew, the "man alive" was not just a cliché but made a real distinction between those who *could* get up and those who couldn't. What I wasn't ready for was how pointedly this scene was about *my* mortality.

But of course it wasn't just about me. Matt was returning my message to seize the day, and it was hardly his fault that I was closer to sleep than to waking. My reaction was not the cool, classical one Housman so often versifies. The moment rattled me in precisely the same way E. B. White reports being rattled at the end of his best-known essay "Once More to the Lake". White returns with his son to a lake camp where his family vacationed years before, remembering how his (now presumably dead) father exercised such authority in unloading the family's trunks at the beginning of their vacation. Little seems to have changed; it is the same lake, the same cabins, the same road to the farmhouse where they took their meals, what even seemed the same dragonfly settling at the end of the boy's fishing pole as he and his son sat fishing, so that the narrator begins to confuse the two: "I began to sustain the illusion that he was I, and therefore, by simple transposition, that I was my father". The same oppressive summer thunderstorm builds, breaks — and then the swimmers go into the water, the boy puts on his already wet swimming trunks, and White's closing line is: "As he buckled the swollen belt, suddenly my groin felt the chill of death."

211

2.

Embedded in one of E. B. White's *New Yorker* reports about wintering in Florida is a three-page gem of an essay. White called the whole piece "The Ring of Time", a title that applies especially to its little, detachable essay about a young barefoot circus rider who brushes past him in a circus big tent one afternoon, then swings herself onto the back of a horse that canters around the ring. The strap of the bathing suit she wears has broken, and she proceeds to roll it up and tuck it into the suit top while the horse does two circuits. White contrasts the nonchalance of her rehearsal with the performance to come in the evening, and the moment of his watching with the moment in the girl's life: she is not so young that she doesn't appreciate "a perfectly behaved body and the fun of using it to do a trick most people can't do," but too young "to know that time does not really move in a circle at all" and that she is older each time she completes a turn around the circus ring.

"The Ring of Time" is White's master theme, obvious after reading only three or four of his essays or casual pieces, from the seasonal observations about his farm in Maine to the poignant pages of "Once More to the Lake". Partly the theme has to do with rescuing from time the fleeting moment that begs to be remembered. "As a writing man, or secretary," White comments of the circus rider scene above, "I have always felt charged with the safekeeping of all unexpected items of worldly or unworldly enchantment, as though I might be held personally responsible if even a small one were to be lost." But the deeper theme is always mortality, the reason enchanting moments are so poignant. Yet White never shoves a skeleton at us; he never leaves it at mortality alone. In "Once More to the Lake", the lake still full of swimmers and the boy going in — that is, the continuum — is as important as the aperçu of the aging father.

3.

One of the ways I hope an essay will develop is by accretion, a growth around a seed crystal provided by a remembered happening. I was sure the scene in which Matt capped my verse would lead to something. I wrote out a little description of it and waited. Often I will begin with such a scene or idea and then other notes or little narratives will seem to be connected and I will put them together in a gradually expanding file. But here nothing happened. I would occasionally look back through my notes and see my description, but nothing came to join it. Of course I knew perfectly well what I wanted to happen. I'd hoped the incident would make my pen go racing, that it would become the center or the opening or the climax of Michael Cohen's version of "Once More to the Lake".

4.

In "The Flocks We Watch by Night", from the November, 1939, *New Yorker*, White and his neighbor wrestle with a ewe in order to get a rope on her and take her back to the neighbor's pasture. They talk about the ram they have bought for their ewes, and when they get the ewe secured, with the help of White's son, who is called only "the boy", the asthmatic neighbor, recovering from the breathlessness of the tussle, shows pictures of his grown children.

The boy asks, as he and White walk home, "What did he mean when he asked you what was going to happen?" White replies that he meant the war. "Do people have to fight whether they want to or not?" asks the boy, and White replies, "Some of them." They walk on, and as they reach White's pasture with its sheep grazing, the boy says, "I can hardly wait to see the lambs."

Again and again White, seemingly effortlessly, assembles his figures representing the Ages of Man (though sometimes it's the ages of sheep or pigs), touches the picture lightly with birth and renewal as well as inevitable death, and says, "There you have it." Or he gives it all to you in a line: "a little girl, returning from the grave of a pet bird and leaning with her elbows on the windowsill, inhaling the unfamiliar draught of death, suddenly seeing herself as part of the complete story." ("Freedom", July, 1940.) Or again, "The worm fattens on the apple, the young goose fattens on the wormy fruit, the man fattens on the young goose, the worm awaits the man" ("Cold Weather", January, 1943). White does not tell you how a king may go a progress through the guts of a beggar, because his interest is not in kings. He does sometimes echo Shakespeare, but I will go out on a limb here and say there is nothing in Shakespeare quite like that moment at the end of "Once More to the Lake". Shakespeare's interest in the theme that is White's specialty is formal, showing itself in the structure of his comedies — a structure not even original with him — or in set pieces like the Ages of Man or Hamlet's reminders to Claudius ("a king may go a progress through the guts of a beggar") and Gertrude ("get you to my lady's chamber and tell her, she may paint an inch thick, to this favor she must come.") White's theme does not lay the emphasis on the end, but juxtaposes an older consciousness with the youth and grace that he puts at center stage.

5.

When I think back now, I can see there were other episodes with my sons that I saw through an emotive lens ground by White. For instance, when Matt was about seven, we sailed across Kentucky Lake for a fishing and camping expedition together. The first afternoon he corrected me when I called a Daddy-Long-Legs a spider. "It's called a Harvestman and it isn't a spider," he said. I was about to point out that it has eight legs and therefore must be a spider when I remembered a little argument I'd had with my stepfather about what happened to the tails of tadpoles as they grew into frogs. He said they dropped off; I said they were gradually absorbed. I was right. And I knew, remembering that moment, that Matt was right, too. I was a little older than Matt when I scored my point, but the fact that he was younger made the moment more pointed for me ("I began to sustain the illusion that he was I, and therefore, by simple transposition, that I was my father").

6.

Roger Angell has a chapter, "Andy", about E. B. White in his memoir, *Let Me Finish*. E. B. White married Angell's mother, Katharine Angell White, after she divorced her first husband. Angell says of E. B. White that he was "the most charming man I've known", and that he got that charm into his writing.

Angell picks a seemingly innocuous passage about driving into Maine on U. S. Highway 1 to discuss, noting "its simplicity and complexity." White had written, "You can certainly learn to spell 'moccasin' while driving into Maine, and there is often little else to do except steer and avoid death." Angell points out that White works the word *death* into the passage as a "sentence-closer", and goes on to say that "the D-word also ends two famous essays of his, 'Once More to the Lake' and 'Here Is New York.'" He decides that "writing about death became a strength for him" that shows up in *Stuart Little* and especially in *Charlotte's Web*, "his masterpiece".

In *Charlotte's Web*, the farmer's daughter, Fern, pleads with her father not to kill Wilbur, the runt pig of the litter. She asks, not in the name of mercy, but justice; "Fern was up at daylight, trying to rid the world of injustice," says her father. Of course, the book is a children's story and at least partly about the deferral of death by magic. The pig's spider friend Charlotte astounds all the humans with the messages she writes in her web: "Some Pig... Terrific... Radiant ," and, finally, "Humble". These signs keep the farmer from slaughtering Wilbur, who lives to a ripe old age. But as Angell notes, *Charlotte's Web* doesn't pretend that death can be kept away. It is the spider herself who dies, alone. Consolation comes in the spring when hundreds of Charlotte's eggs hatch into spiders and either stay on the farm or go their ways through the world. The renewal of life is the closest thing to justice in the cosmic scheme, but death is its price.

7.

"Morningtime and Eveningtime" appeared in White's October, 1942, "One Man's Meat" column in *Harper's*. White juxtaposes a rural civil defense airplane spotting post with "the rich black afternoon" of a pasture and henyard, "a little portion of America, imperiled, smiling, and beautiful as anything." Another contrast is this smiling countryside with "over there", imagined as somewhere in the European Theater of the war. As White finishes his watch, his phone rings to say the yellow alert has been changed to red. He picks up

> the ewe whose lamb we butchered in warm blood day
> before yesterday... the same lamb I sat up late in March

for, with such apparent tenderness, the kind of tenderness
which refuses to look ahead because it knows that to look
ahead is fatal. The tenderness of March, the brutality of
August, one lamb serving both moods.

The tenderness of March, the brutality of August,
the guns of September and possibly the enemy planes
of October. And for the lamb no sign, "Some Lamb".
White drives the country road, giving the blackout
signal to his neighbors, and then returns to his house.
"Here, in the compulsory dark, I sit and feel again the
matchless circle of the hours, the endless circle Porgy
meant when he sang to Bess: 'Mornin'time and eve-
nin'time...'" At the end of the column White sits down
at his typewriter and tries, without success, to write,
but we know that the time for that will come around
again, too.

8.

My last year of teaching before retirement was also
Matt's first year of teaching. He teaches literature, as I
did for thirty years. All during that year I had a very
poignant sense of handing over to him, passing on to
him, not exactly the family business, but the endeav-
or, the calling, the idea that what we do matters, that
trying to teach people about language and art is worth
spending our lives doing. Always sensitive to what I
was feeling, he called me up every once and a while
to ask me questions. "I'm starting drama today; what
should I say?" He made me feel as if I had something
worthwhile to pass on. But of course, his attentiveness
also made me aware that I was passing on. Well, we
can hardly fault our children for doing what we hoped
they would do, for carrying on.

20. On the Road Again

"A swift carriage, of a dark night, rattling with four horses over roads that one can't see--that's my idea of happiness."

  – Isabel Archer, in Henry James's *The Portrait of a Lady* (1881)

I don't like driving as much as my friends and family say they do. I'm happy to let others take the wheel, and I often think while driving about the things I would rather be doing. But this reaction is mere churlishness on my part, because driving is the Great American Birthright and long ago replaced baseball as the National Pastime. We Americans affirm our passion for driving by burning twenty million barrels of oil per day, a quarter as much as the whole world consumes.

Recently, I decided to reread Jack Kerouac's *On the Road* to try to recapture some of the feeling being on the road used to have for me. It took me a few days of searching used book shops to find a copy that was beat enough. Eventually I found one so worn the pages were disintegrating as I finished reading them. My son Matt, visiting for the week, noticed I was reading it.

"Didn't you read it back in the day?" he asked.

"You mean the sixties?" I said. "I was a little young for it when it came out in 1957. I was only fourteen."

"Whenever," he said.

"You know I'm not sure. I was pretty sure that I'd read it in college, but not much of it seems familiar to me."

"Why is it," Matt asked, "that whenever I ask someone about the sixties I get that kind of answer?"

This was a rhetorical question, I decided, and didn't need an answer.

Truly, Kerouac's book did not seem like familiar territory to me. I read through Part One, in which Sal Paradise meets Dean Moriarty and they both travel, but separately, around the country. They are hitchhiking and encounter each other briefly in Denver. Neither has a car of his own, and the only real driving that's done by Dean Moriarty is in the mountains above Denver, where he steals a car and drives back down the mountain to Denver at ninety miles an hour.

I began to muse about what it was like to drive at the age of twenty. Did the early appeal of driving for me have to do with speed and danger? At least I can say for sure that as driving has gotten more comfortable and safer, I have grown out of fondness for it.

II.

When I was in college I had my first car, a 1960 Ford Falcon. My drive home every few weekends was a two-hour transition from the high desert of Tucson to the low desert of the Salt River, through tiny desert towns. Picacho was hardly more than a few trailers and a gas station, but memorable because of the strange little mountain there, Picacho Peak, visible from the road seventy miles away as I rolled through the Santa Cruz Flats, a basin area untypical of generally mountainous Arizona. Another town, slightly larger and once a railroad stop but by this time dying, was Eloy. ("Aptly named," said my priest friend, John Fahey, "after that despairing cry of Christ on the cross — Eloi, Eloi, lama sabachthani!") Farther north, around Casa Grande, some concrete slabs half-covered in sand were all that remained of the Japanese internment camps of WWII. Finally I would come to the cotton fields of Maricopa County and the little town of Chandler, where I had gone to high school. I drove fast, much faster than was reasonable or prudent. The Arizona speed limit, as the drivers' manual said in those days, was "Reasonable and Prudent" when not posted, and those words were traditionally given a liberal interpretation by Arizona drivers, if not by Arizona cops.

Then my parents moved to California. After their move, on my less frequent visits, I would head west to the Arizona-California border at Blythe and then on through the Mojave, through Bakersfield and onto the old San Bernardino Freeway, the first one built in the country. Gradually the traffic would speed up, even from the highway speeds of the desert west.

The sensation of speed, by itself, does not seem to me in retrospect to have been a big pleasure of mine. But, what got my blood pumping, I do recall, was *driving fast while everyone around me was close and also driving fast*. On the approaches to Los Angeles on the San Bernardino Freeway, the old freeway was level and mostly straight. Gradually, as I neared the Hollywood Hills and the eventual merging with the Santa Monica Freeway — I was going to Santa Monica, where my parents lived — the traffic would increase, and eventually I would be surrounded by cars, all going faster than seventy miles an hour. Then the roadway would lift and turn as we approached the interchange where we crossed the old Hollywood Freeway as it turned into the Santa Ana Freeway. The formation of cars around me would lift and turn as well. In *The Outermost House*, Henry Beston talks about flights of shorebirds that "rise as one, coast as one, tilt their dozen bodies as one, and as one wheel off on the course which the new group will has determined." Our freeway flight also seemed to have a group will — the will of the roadway, at whose bidding we tilted and turned.

As it approaches the great East Los Angeles Interchange, the San Bernardino Freeway passes over, under, and through a series of connecting ramps that shuttle thousands of cars an hour among eight huge arteries of the most extended city in the country: the Golden Gate Freeway that comes down from the San Fernando Valley and hints at the whole expanse of northern California above it; the Hollywood Freeway, with its suggestion of the "real" Los Angeles; the Harbor Freeway that takes you straight south to the shipyards of Long Beach; the Long Beach Freeway that sends you on a more leisurely trajectory to the more touristic parts of the south coast; the Santa Ana Freeway that projects you down into Anaheim and Disneyland, which somehow models all of Los Angeles — what Kerouac called "the whole mad thing, the ragged promised land, the fantastic end of America" — in cleaner pastels and rides that only slightly exaggerate the highway cloverleaf swoops. Finally there was the Pasadena Freeway that ended practically at the door of the Huntington Library in posh San Marino, and the road I was looking for — the Santa Monica Freeway, a straight shot west to the beach. At this point, my own formation of cars sped through similar formations whizzing to right and left, overhead and below. At the point where the freeways cross there are seven levels of independent, blurring speed in all directions through the air. Driving at speed on this interchange bears the same relation to ordinary driving as landing a fighter jet on an aircraft carrier must to sitting in the economy cabin of a taxiing commuter airliner. At the speed my 1960 Falcon and the cars around it were moving, a blunder probably would have meant an unsurvivable crash, without seat belts, with hard dashboards, in cars whose steering columns did not collapse — a safety engineer's nightmare of steel that was rigid where it should give and thin-pressed sheet where it should rigidly enclose. Of course I didn't think about crashes. I loved it. I felt jolted awake as if

221

from a drugged sleep inspired by four hundred miles of straight and level, nearly featureless desert driving.

III.

As we find out in Part Two of *On the Road*, Kerouac's narrator Sal Paradise hates to drive and doesn't even have a license. He drives carefully, he tells us, but Dean Moriarty makes up for Sal's caution with his own speed and recklessness. By this time Dean has acquired a huge brand-new Hudson, whose bench front seat easily accommodates the four people who are often riding there.

I got my first driver's license the year *On the Road* was published. I was fourteen years old. Because of a perverse Arizona law that put the most dangerous vehicles in the hands of the youngest drivers, my license was only valid for motorcycles and scooters like the Vespa I drove to and from high school. It was 1957, the year I started high school. That was also the year work on the Interstate highway system began.

According to Tom Lewis' 1997 book, *Divided Highways*, a history of the Interstate system, around the time I was born in 1943, the number of highway miles in the United States passed the number of miles of train tracks. The Depression had spurred federal work projects to build state roads and highways. The Pennsylvania Turnpike opened in 1940; the New York Thruway was begun in 1948; the New Jersey Turnpike opened in 1952. But it became clear that these individual state efforts would never catch up to the national need for good roads for the hugely-expanding car travel of the post-war period. In the 1956 law passed at the end of Eisenhower's first administration, the Interstates were supposed to be built in 13 years; it took 40. Eventually there would be more than 40,000 miles of interstate, each with access only by ramps, at least four lanes, a concrete roadbed, gradual grades, and turns so contoured that they can be taken at speeds of seventy miles an hour or faster.

Few people have written about the aesthetics of freeway driving, though Joan Didion's *Play It as It Lays* comes to mind. Generally the charm of American highways is a wholly nostalgic one, like our strange obsession with Route 66, or is imagined to lie in the smaller roads, the "blue highways" on the map that William Least Heat Moon traveled for his book of that name. But in an odd book called simply *Roads*, Larry McMurtry makes the case for interstate driving. I call his book odd because McMurtry's method seems to be to think about or describe places as he *drives by them*. The book is full of sentences like "I also passed up a chance to revisit -----." "I wanted to drive the American roads at the century's end, to look at the country again, from border to border and beach to beach," McMurtry writes, and so he does, often driving 700 miles a day across the country on Interstates 10, 40, 70, 80, or 90 or down its length on 5, 25, 35, or 75. Not interested in poking into strange places, he drives "the great roads" as he calls them, "whose aim is to move you, not educate you." He shares my prejudices for warmth and open spaces, preferring the south to the north and the west to the east, refusing to drive on I-95.

McMurtry is old enough to appreciate the difference the interstate roads made in enabling us to get around this huge country quickly. Getting there is the point of the road, McMurtry suggests. Though they have adventures along the way and travel before interstates, Sal Paradise and Dean Moriarty also have this attitude about roads. They are always looking for the "final" cities, the end of everything at the end of the road: New York, Los Angeles, San Francisco, and finally Mexico

City. Unlike McMurtry, they hardly even talk about what they pass by, so hurried are they to get to the end.

IV.

"He gave a flick to the horses, / and with streaming manes / they ran for the open country" recites Derek Jacobi on the audiotape in my car. It is Book Three of the *Odyssey*, the original on-the-road story. I am driving east toward the rising sun in southwest Missouri on Highway 60, a road so narrow, winding, and old that it is all being replaced. I have just been visiting my mother in Springfield, and I am on my way back to Kentucky, where tomorrow I will be teaching the *Odyssey* to two humanities classes. I will also be chairing the English department at my college.

When I took on the chairman's job, I found I had little time left for reading and class preparation. Recorded books saved me during my tenure as chairman; I listened to them on the drive separating home and work, and the six-hour drive to my mother's house every few weeks let me hear whole books. I listened to Seamus Heaney read his own translation of the old English epic *Beowulf* on that route. *Beowulf* begins and ends with a voyage. In fact, many of the works of the European and American literary heritage are on-the-road stories: *Don Quixote*, Fielding's *Joseph Andrews* and *Tom Jones*, Byron's *Childe Harold's Pilgrimage*, Mark Twain's *Huckleberry Finn*.

At the moment it is Book Three of the *Odyssey*, in which Odysseus' son Telemakhos rides in Nestor's beautiful inlaid chariot, yoked to swift, willing horses, accompanied by Nestor's son Peisistratos. They rush from dawn to dusk for two days from Nestor's palace in sandy Pylos to Lacedemon and the hall of Menelaos. Later they return, in an almost identical passage, at the identical speed. It is a young man's pace, and Homer's lines assure us that Telemakhos is enjoying himself. Telemakhos doesn't do any of the driving; like Sal Paradise in Part Three of *On the Road*, he leaves the driving to his young friend. Dean Moriarty drives a travel-bureau Cadillac from Denver to Chicago without stopping; Peisistratos drives the chariot from dawn to dusk without pause for two days. Meanwhile, Odysseus' own travels continue, but there is nothing to indicate he feels the pleasure of traveling for its own sake. He doesn't want to be on the road; he wants to be back in Ithaca in his own hall with his wife. "There is no boon in life more sweet, I say, / than when a summer joy holds the realm, and banqueters sit listening to a harper / in a great hall, by rows of tables heaped / with bread and roast meat." So Odysseus says in the hall of his host, Alkinous, but it's his own hall he's thinking about.

The Missouri road I am on is traversed with cracks badly filled with asphalt. Every half second the wheels jolt over the cracks, first the front wheels, then the back — klunk-klunk! Being on a bad road ought to be a metaphor for more important things. One has a feeling of not being able to escape, of being trapped, of having to go down this road, because of course one must: this is the road one is on; there aren't alternatives that go to the same place. After a while the steady jolting where the asphalt has been slopped across the road to fill separations becomes routine and I anticipate the regular rhythmic jolts, but every once in a while a new and bigger one will suddenly surprise me. I move out into the passing lane because it offers the illusion of escaping the worst of the bumps, but then I discover that lane has some of the same horrors and a few special potholes of its own. I can speed up, of course. That won't lessen the nastiness of the road, but it will add the element of danger.

Meanwhile Telemakhos has made it back to Ithaca, warned by Athena of the dangers ahead and so escaping the ambush planned for him by his mother's suitors, evil men who are...

A car changing lanes ahead brings my attention back to the road, and I begin to muse on how complicated is the notion of attending to the act of driving. "Paying attention" is at all times a very selective process. We constantly filter the information from what we hear, see smell, taste, and, touch, or we would suffer a sensory overload. The mind that takes in everything is not the normal mind, but the autistic or obsessive-compulsive one. What you do pay attention to is the road, but "watching the road" is already an abstraction because you don't watch the road as a whole but portions of it, coming at you and receding from you in the rearview mirror. You don't watch the road as far as you can see (unless you are driving at night) or directly in front of the car, but a middle distance, and even that you don't watch continually, because "watching the road" means being aware of what might come onto the road from the sides: cars from side streets or deer from woods and ditches.

Where is your attention during those moments when you are thinking about something other than what is before your eyes? When I'm driving and listening to a recorded book on the tape player, I sometimes suddenly find myself aware of having missed several minutes of the tape. My full attention was commanded by something on the road, this truck pulling into my lane in front of me, for instance. That need for my full attention may last only a few seconds, but my attention stays on the road exclusively for some minutes, until I suddenly became aware again of what I was listening to and that I have missed several minutes.

The converse experience is driving and suddenly

realizing that I have traveled some distance, perhaps a couple of miles, without being aware of the road at all, having been absorbed in listening to a tape or the radio or simply in daydreaming. Maybe a mile, maybe more. But I must have been watching the road; I have to assume I would notice a dangerous situation and react to it, because I *have* come out of such reveries to brake at an obstruction or to react to a car I was approaching too quickly. I could come back to full road attention at a signal alarm stimulus like that, but in such a reverie I would have missed a turn I had been intending to take. My thoughts were somewhere else, and I couldn't tell you what roads I had passed, or stores or houses. What does it mean not to be attentive in this way? How many miles have I driven not really paying attention to the road or the passing scene?

V.

"We know time," Dean Moriarty keeps saying in *On the Road*; it's his mantra. He means that we know time is passing and the moment must be seized and lived: "we know time — how to slow it up and walk and dig." But this version of his refrain he speaks standing in the rain, as Sal Paradise is about to go west, leaving Moriarty behind in New York. "Damn! But the mere thought of crossing that awful continent again," says Dean, not finishing the thought. The road is the master metaphor for life and the road goes on, whether we are on it or not. At points in *On the Road*, the American road becomes both too much and not enough. It is too much in a scene late in Part Two where our travelers — Sal Paradise, Dean Moriarty and Dean's wife MaryLou — are driving through the swamps of western Louisiana. These hep roadsters who usually dig everything about the American scene are suddenly freaked out by this part of America, so much so that they reject it as not America at all. "We were surrounded by a great forest of viny trees," writes Sal, "in which we could almost hear the slither of a million copperheads." Suddenly there is an "apparition" of a black man by the side of the road with his arms raised "praying or calling down a curse." They gun the Hudson and don't slow down until they reach the polluted air of the refineries outside Beaumont, Texas, where they can relax because they're back in America. "We wanted to get out of this mansion of the snake, this mireful drooping dark, and zoom on back to familiar American ground and cow-towns." But the American road is too little in Part Four, as Dean and Sal drive in a '37 Ford through mountains and jungles into Mexico, encountering marijuana cigarettes as big as cigars and whorehouses full of teenaged girls. Mexico City, where they are headed, is the dream city of this barely post-adolescent pair, "the great and final wild uninhibited Fellahin-childlike city that we knew we would find at the end of the road."

VI.

During this last Christmas season we managed to get the whole family together. I was driving my two sons — both in their thirties — in from the airport when Matt, the elder, began to criticize my driving: I was following too close, I was going too fast, I was running yellow lights turning to red. Dan, the younger boy, nodded in agreement. The next time we got in the car I handed Matt the keys. It never bothers me to hand over the reins — I don't feel the need to be in control whenever I get in a car.

I sat on the passenger's side and remarked, with hubris that now makes me wince, that I'd been driving for forty-six years and never had an accident, while the two boys had already had two each. They were indignant. "But those were not our fault," they said, with perfect truth. Yet the actuarial tables don't make this distinction between the innocent and those at fault: young men, especially between eighteen and twenty-five, have more accidents—*have* here being a term neutral as far as assigning blame—than women in that age group or than men and women in any other age group.

I began to think back over those forty-six years and what I had avoided. What, I wondered, had the chances been of my being killed during all that driving?

I started driving at fourteen. My Vespa motor scooter took me all over the area south of Phoenix where we lived. No one wore helmets; in those days the proposed law mandating their use, ultimately voted down in Arizona's libertarian legislature, had not even been thought of. When I left for college in Tucson I had no car, but I soon acquired my stepfather's 1960 Ford Falcon. In graduate school I had a used Peugeot 404, eventually destroyed by a drunk driver as it was parked in front of my house in New Orleans. I drove my wife's Chevy Malibu for a while. In 1974, with two young boys now, we bought an American Motors Hornet Station Wagon, the only new car I've ever owned. Later I drove a 1979 Lincoln Town Car, another hand-me-down, and a 1992 Camry bought used. Each of these last two I drove more than two hundred thousand miles. Other people's cars that I have driven for shorter distances, from a few hundred to a few thousand miles, include an old Volkswagen Bug, a Volvo Sedan, and a Plymouth Voyager.

As nearly as I can figure it, I've probably driven more than 700,000 miles in those forty-six years. Coincidentally, that's about the number of miles of highway in the country built with the aid of Federal dollars, in-

229

cluding the forty thousand miles of interstates. Almost three quarters of a million miles.

According to California's Department of Motor Vehicles, drivers from the ages of thirty to sixty are at their lowest risk of having a fatal accident, which happens in this age group about once every million miles. After the middle sixties, the chances start going back up. Teen-aged drivers are about four times as likely to have a fatal accident as those in the thirty to sixty age group. If a teenager drives as much as a hundred thousand miles, he or she has about a fifty-fifty chance of being involved in a fatal accident. Twenty-five year old drivers are about twice as likely to die in an auto accident, given the same number of miles driven, as their older counterparts from thirty to sixty. I worry about my sons on the road, though not nearly as much as I did when they were in their teens and not yet home at two in the morning. But I still worry and try not to think about the road's dangers. But I have it backwards, I reckon. They are right to worry about me—they are coming out of the high accident age. I am going back in.

VII.

Such a project as the interstate system was an expensive and monumental undertaking, requiring 200 feet of right-of-way and many buildings' worth of poured concrete for each mile. Eisenhower was convinced of the need, but he had to convince the Congress and the country. He did it with national defense and the Cold War: the interstate system was to be used not only to transport military supplies and troops quickly across the country, but its wide roads would be escape routes from population centers in the event of nuclear attack.

As *On the Road* draws to its end, Sal Paradise for the first time mentions The Bomb, the other child of the fifties growing up in the sixties, the ultimate Baby Boomer. He says the peasants of the rural Mexico they're driving through have no idea that now there is a bomb that can destroy all of them. Hearkening back to a time before the bomb is a part of the book's elegiac tone; all the adventures Sal is relating are themselves now in the past, Sal has settled down, and Dean is gone from his life.

During the sixties I remember two recurring dreams. In the first I am driving to the north, from Tucson back to my parents' home (though by that time they had already moved to California) and ahead I see the mushroom cloud of a nuclear explosion. There is nothing to be done. I turn and drive the other way.

The other dream I still have occasionally. In my dream I am still in my twenties. It is in the west, somewhere, probably Arizona. I have a car—I don't know what kind of car. I apparently have money enough that it does not figure in my dream—what does figure is *time* and the *road*. I am free from school or work for some weeks, perhaps months. I realize I can go in any direction I choose. I can go south to stay with Dave in Tucson or Pat in Ajo. I can go west to see friends on the coast or just wander in California. I can go north to the cool mountains or the Grand Canyon, or east to the whole rest of the country. The road is all before me. I also have a sense that I will not take real advantage of this freedom.

21. The King in Winter

"It doesn't look much like a teapot to me," said my wife as we stood gazing at the constellation of Sagittarius in the southern sky. "And it certainly doesn't look like an archer. I think it looks sort of like a golfer with one foot in a sand trap."

I couldn't see the golfer, but I certainly saw the problem. Few of the constellations look much like they are "supposed" to — that is, as the Greeks and Romans saw them and named them. Leo is a spectacular exception, with its curve of stars arching up from Algieba and defining the maned head of the lion and its hindquarters the triangle of stars ending in Denebola on the trailing or eastward side. Cygnus looks enough like a swan to be plausible, if you can make out the faint double star Albireo marking its head. And Aquila with a little imagination can be made to seem like an eagle flying by Cygnus on an opposite course along the Milky Way. When you can see the whole of Scorpius — which is rare above 35° of latitude — it does bear some resemblance to a scorpion, but Draco looks more like a snake than a dragon, and the Ursas, major and minor, are not, let's face it, very ursine in shape. The arbitrariness of the shapes people have seen in the stars is perhaps best illustrated by the part of Ursa Major that Americans see as a big dipper and Europeans see as a plow.

H. A. Rey's 1952 book *The Stars: A New Way to See Them* introduced a simplified way of connecting the major stars in each constellation that was so effective it has been adopted by many star guides, including *Sky & Telescope*'s monthly Sky Chart and the magazine's illustrated observing articles. But aside from the stick figures of the Twins and Virgo and the ones already mentioned, Rey's diagrams don't really *animate* constellation shapes. The connected stars show relatively abstract shapes in the sky. The observer still has to provide the imagination.

My son Dan did that for me one winter night when we were looking at Orion. "I don't know how people see a hunter there," he said. "It seems obvious to me what pattern those stars make," he said, striking a pose with his right arm high in the air, left hand off to the side, hips tilted and knees flexed. "It's Elvis!" I said. "How could I have missed it before?" Perhaps the fact that Dan makes his living as a guitarist and songwriter affects the way he looks at the stars, but I know I'll never see that constellation the same way again. The next time you're out in the winter dark on a clear night, see if you can't find Elvis, his tilted pelvis accented by a rhinestone belt and Betelgeuse even providing a red "on" light in the microphone he's holding in his right hand.

22. My Next Read

It would be worth the while to select our reading, for books are
the society we keep...

> – Thoreau, *A Week on the Concord and Merrimack Rivers* (1849)

I sometimes have trouble choosing the next book to
read. Plenty of advice is available for this particular
form of abulia. "Choose an author as you choose a
friend," writes the fourth Earl of Roscommon. "When
a new book is published, choose an old one," advises
the sententious Samuel Rogers. For more up-to-date
help, since these gentlemen were writing in the seventeenth and eighteenth centuries, type "What should
I read next?" into a search engine. Half a dozen website addresses will pop up; their method is to ask you
what you've read in the past and how you've liked it,
and then to make recommendations according to an
algorithm like the taste analysis tools that are part of
Amazon.com's success. Whenever I buy a book online,
Amazon tells me what other books were bought by the
customers who last ordered my title, and whenever
I go to the Amazon site, I get suggestions of books I
might like. Despite their sophistication, these recom-
mendation procedures, which the techies call "collab-
orative filtering", have yet to show me anything that
excites my interest.

Book clubs are another source of help to the indecisive reader. There are even online book clubs, though
I think the idea of a book club ought to be to temper
the solitary experience of reading with the opportunity
to discuss it with real people, present, in the flesh. But
book clubs *assign* books; I want to *choose* one.

And in any case my problem is never that I cannot think of something I'd like to read. Most of the time I have a huge backlog of books I'd like to read, recommendations from friends, titles I've recently been reminded of that I'd always meant to read, and often, as is true just now, an actual shelf of books that I have merely to pick up, open, and begin reading. Part of the problem of what to read next is this question: why is it more difficult than just picking one up?

Recently, I started to read *The Charterhouse of Parma*, but I got stalled between the details of Napoleon's two Italian campaigns and the tone of the book, which is unrelenting sarcasm toward the petty nobility of Lombardy. About the same time, I read and liked the first chapter of Howard Zinn's *A People's History of the United States*, but I don't seem to have read beyond it. I know well that reading often requires getting beyond the first few dozen pages and a first impression. But most readers, I think, will understand me when I say that a book can have a presence that is dense and resistant.

The Resistance of the Text

The first time I tried to read *Howard's End*, I threw the book down in exasperation after fifty pages. A year or so later I picked it up again and found it enjoyable right through to the end. I have since read a lot of Forster, but there is an adjustment I have to make when I begin reading. My problem, I suppose, is that I read from the inside: I have to be a sympathetic rather than an aloof or defensive reader. I like to become the reader I sense the writer's looking for and at least understand if not wholly assent to the writer's view of the world. With Forster, that view is apparent in every paragraph — a sort of awful sincerity, a sentimentality about the lower classes that seems almost the product of a guilty conscience, and of course, the unrelenting theme: "Only Connect". With my more recent experience with Stendhal, I had to contend with my own ignorance of events from the occupation of Milan to the Battle of Marengo and beyond, but also with Stendhal's concentration on Fabrizio's father, whom the author did not like and who was not going to be important in the whole narrative.

Then I started Walker Percy's *The Moviegoer*. Here I was on familiar ground, since I had lived in New Orleans, and not too far from the Gentilly location near Chef Menteur where the narrator has exiled himself. But Percy's complicated mind kept me from relaxing into the book initially.

So for a while I was bouncing among several books — Percy, Stendhal... oh, and I had also picked up a nonfiction book, Nathaniel Philbrick's *In the Heart of the Sea*, about the wreck of the whale ship *Essex*, an event which inspired Melville when he began thinking about *Moby Dick*. This book was sitting on my to-read shelf and had been given to me by my younger son a few Christmases or birthdays ago. I read the introduction and the first chapter, but it seemed to move slowly and there was a good deal of what "the cabin boy would have seen in the streets of Nantucket in July of 1819." So I put it aside. I went back to Stendhal and began chapter three: "Fabrizio soon encountered some canteen-women, and his extreme gratitude to the jailer's wife" — wait, Fabrizio had been in jail? After a few days I had lost the thread of the story and had to backtrack to the middle of the previous chapter. Admittedly, the beginning of a book can require special attention, but I knew I had in my hands the sort of book that, once taken up, requires that one read some of it every day. Any novel with an interesting narrative line, let alone a complex one, will require such attention, and reward it, but Stendhal's Fabrizio was competing for my next read, and he lost before he even made it to Waterloo.

I picked up *The Moviegoer* again and read it through, turning to *In the Heart of the Sea* only occasionally for a break. The *Essex* story had become more interesting once Philbrick launched his boat and crew from Nantucket.

When I finished *The Moviegoer* what might have happened was a default choice: I had another Walker Percy, *Love in the Ruins*, on my shelf. Defaulting to the same author whose book has just engaged me is a choice I made more often when I was younger. I frequently would stay with an author for two, three or more books, sometimes for an entire oeuvre. I went through Patrick O'Brian's Aubrey/Maturin series with hardly a break when I discovered these books ten years ago.

In recent years the only writer who has captured me for more than a second book is Penelope Fitzgerald. After reading *Offshore*, her novel about a group of people living in leaky barges on the Battersea Reach of the Thames, I was intrigued enough to pick up *The Bookshop*, about a widow who decides to open a bookshop in the aptly-named town of Hardborough. She is opposed passively by a town apathetic to books and actively by Mrs. Gamart, the local Lucia, who has her own plans for the building the bookshop is in. The widow perseveres, and for a while she prospers. The book is often funny; the local power struggles recall E. F. Benson, and although I thought of Muriel Spark and Barbara Pym, Fitzgerald's voice is unique, and as I found by reading further, she never repeats herself. I discovered she had written a mystery, *The Golden Child*, a delightful story, part mystery and part social comedy, which takes place in a museum that is not quite the British Museum during an exhibition that is not quite that of the Tutankhamen artifacts. *Human Voices* is a novel about Fitzgerald's experiences at the BBC during WWII. She also wrote a few short stories, and these too, are so unlike each other that they might have been written by different people, but for her distinct perception and her clear, spare style—a succinctness that means her books tend to be quite short. The familiar within the various satisfied for me that paradoxical longing that impels us readers to seek out another book by an author we've enjoyed. We want more of the same, but of course we don't really want it to be the same. We want the same goodness rather than any other sameness—the goodness being style or an approach to experience we find appealing or some other indefinable feature that makes an author unique.

But I did not pick up *Love in the Ruins*. Instead I vacillated for a while between Dee Brown's *Bury My Heart at Wounded Knee* and William Golding's *Rites of Passage*. To stop the vacillation, I read right through one of Margaret Maron's Christmas mysteries. I have less trouble getting into quicker reads like mysteries. Genre fiction does not usually offer the kind of back pressure to the reader that mainstream fiction can. Indeed, quick entry into the action and the presentation of a character who arouses immediate curiosity are defining features of classic mysteries, westerns, and probably science fiction and romances as well. Look at the first page of Raymond Chandler's *The Big Sleep* or *Farewell, My Lovely* to see how fast we are made eager to find out the issue of the narrator's interview with "four million dollars" or why the huge, ham-fisted man ("but not more than six feet five inches tall and not wider than a beer truck") is gazing up so mournfully at the windows of the "dine and dice emporium" called Florian's. After the second paragraph of *Riders of the Purple Sage* we want to know what Jane Withersteen's story is, why she dreads the impending meeting with the church elders, and who the "Gentile" is she has befriended. Sometimes the first sentence is enough to capture us; the whole theme of Dashiell Hammett's *Red Harvest* is anticipated in its opening: "I first heard Personville called Poisonville by a red-haired mucker named Hickey Dewey in the Big Ship in Butte."

The reader may retort here that the openings of many mainstream novels are striking and memorable.

> "It is a truth universally acknowledged that a single man in possession of a good fortune is in want of a wife."

> "All happy families are alike; every unhappy family is unhappy in its own way."

But these openings don't propel you into a story before you know what's happened; they make you stop and think before you proceed — and proceed with care as well as curiosity.

I remember the first time I tried to listen to an audio recording of an Austen novel in my car. It wasn't *Pride and Prejudice*, quoted above, but *Emma*, which begins: "Emma Woodhouse, handsome, clever, and rich, with a comfortable home and happy disposition, seemed to unite some of the best blessings of existence, and had lived nearly twenty-one years in the world with very little to distress or vex her." I was driving from Kentucky down to New Orleans, and I didn't turn on the tape until I was near the Kentucky border near Fulton. I listened to that first sentence and immediately had to stop the tape and play it back; what exactly about that description conveyed the author's ambiguous feeling toward her heroine? Of course that "with very little to distress or vex her" was part of it: in Austen's world, if one hadn't been vexed in all that time one had either been spoiled or just not paying attention. And when

I played it again the word *clever* jumped out, too. For Austen, words describing powers of mind are rarely neutral: *intelligent* is positive, but *clever* is not, and marks Emma's power of mind as possibly mischievous, even dangerous.

And so it went as I tried to concentrate on traffic traveling the little roads down through Union City, Dyersburg, Millington, and on into Memphis. For the first half hour or so I was punching *stop* and *rewind* often. Finally I forced myself to follow the story without backtracking through Austen's nuanced language.

When I'm listening to a mystery rather than a book like *Emma*, I rarely feel that I have missed anything. Moreover the choosing is different. I review mystery books for my local public radio station, and selecting books here is a different process from that I use for other books. In doing these commentaries, I am not interested in sampling a typical or representative group of mysteries, because that would mean I would have to pan some, even most, of them, and I have no interest in reading bad books, let alone talking about them on your subsidized air time. I scan annotated lists of new book titles put out by specialty shops like Clues Unlimited in Tucson, The Raven Bookstore in Lawrence, Kansas, and Seattle Mystery Bookshop, I shelf-read a good deal in the mystery sections of bookstores, and I rely on other people to give me some idea about what a mystery they've read might be like.

Once I make a choice, getting into the book is rarely much effort, and the reading goes quickly. But I must have more substantial reading along with my mysteries.

The Five-Foot Shelf and the Two-Foot Shelf

Like many readers I've talked to about this, I carry a virtual list in my head of books that I want to read some day. I couldn't recite the list, but when I look at a book title, I know whether it's on my list or not. Some of the books on the list are Harvard Five-Foot Shelf or Mortimer Adler Great Books kinds of classics. Others may be found on those compilations of Most Influential Books that magazines make up from time to time. Others get there because I've seen them mentioned many times in reviews or I've become aware of them as winners of various honors — the National Book Awards (I've read at least a dozen of the winners), the Pulitzer (dozens) or the Man Booker Prize (half a dozen).

My list changes with my age and experience. Years ago, enjoying Plutarch's parallel lives, I thought that all the Greek and Roman historians should be on my list. Now I know that I'm probably not going to read Livy, though I may go back to Suetonius. And the bits and pieces of Herodotus and Thucydides I've read will probably have to last me. But who knows. There may come a time when I will banish all other books and return to spend my dotage in Greece and Rome, Persia and Macedonia. Gibbons' *Decline and Fall* is still on my list, though life may be too short for me to tackle this one. Not that I have stayed away from longer books. One of the first things I did after retiring was to read Marcel Proust's *Remembrance of Things Past*.

My two-foot shelf is not virtual, but actual, and on it I put books that I actually have in hand that I intend to read. Alan Paton's *Cry, the Beloved Country* is an example of a book that has moved from my virtual list — as a book that I heard about in many contexts while I was growing up — to a real shelf in my library, when I recently found an inexpensive, like-new copy for sale. That a booklover buys a book he intends to read is nothing special; half the books in some people's library might fit into that category. But some years ago, I determined to make my library smaller and to try to buy only books I was honestly likely to read, and sooner rather than later. I have followed through by selling or giving away about a thousand books, and the two-foot bookshelf has not yet overflowed.

Longer, Smaller Books

I read the first chapter of Darwin's *The Origin of Species* a few months ago, but I haven't been back to it; it's on my two-foot shelf in a nice Collector's Library Edition, one of those clothbound octavos put out by Barnes and Noble in a size that I like the handling of. There is another essay here on the readability of books that goes beyond their content to size, heft, binding, durability, and feel. Is anyone besides myself willing to admit scanning shelves—in both retail and used bookstores, for the trim, smaller hardbounds that are so pleasing to the hand? Handshopping, I call it. For me, it's liberating, this search for books that doesn't care about authors, subjects, or titles. Of course, these books are likely to have already passed some popularity or time tests to end up in such editions, at least in America. My friend Hughie has often observed that American books tend to be too big. So finding a book that's kinder to the hand than the 8-1/2 x 9-3/4 block that is the staple of the retail American bookstore usually means finding a book that's been around for long enough to be republished in a smaller edition.

Of course, *War and Peace* will not be found in a little volume. When I determined to read it last year, I began searching for a multi-volume edition, hardbound, that would reduce each part to a handy size, and I found it in the Everyman edition. The burgundy buckram-covered volumes were a joy to handle, and I took my time with them, reading twenty pages or so a day, so that I had several months pleasure and the sense of living with Tolstoy's characters over time.

"*War and Peace!*" my friends say. "Proust! Those are big commitments of time. I don't know..." Well, yes, I took a long time reading *War and Peace* — probably more than two months, but I was deliberately limiting the daily reading to savor the book and let it work on me. But we need to look more closely at this "big commitment of time" that is a frequent reaction. What does it mean, exactly? Does it mean that I would be taking time away from what I really wanted to do — rebuild that 1941 Chevy, say, or repaint the barn and then research Pennsylvania hex signs to paint on the front? Probably not. With most people, I think, the time spent on a big reading project is just time that would have been spent on several smaller reading projects. If the book rewards the time — and only you will know that — it's worth the time. I found Proust surprisingly unresistant when I started reading the long first section in which he describes going to bed early as a child — a description that then moves into remarkable observations about the way that thought and memory constitute identity. I knew that at any point in the seven volumes of the book I could have stopped but was led on almost imperceptibly to the end.

Any book is a commitment of my time, as I am well aware here in the middle of my seventh decade. But reading a book is a glorious exercise of free will, from the first opening, through whatever false starts, to the final page. I can always change my mind. A lot of books I have read a page, two pages, fifty pages, and even more, but put down and never picked up again. But tastes change, and one day, next week or next year or sometime, I will pick up *The Charterhouse of Parma* and my interest will never waver through its four hundred pages.